T0254565

SpringerBriefs in Petroleum Geoscience & Engineering

Series editors

Dorrik Stow, Heriot-Watt University, Edinburgh, UK

Mark Bentley, AGR TRACS International Ltd, Aberdeen, UK

Jebraeel Gholinezhad, University of Portsmouth, Portsmouth, Hampshire, UK

Lateef Akanji, King's College, University of Aberdeen, Scotland, UK

Khalik Mohamad Sabil, Heriot-Watt University, Putrajaya, Malaysia

Susan Agar, Houston, USA

Kenichi Soga, Department of Civil and Environmental Engineering, University of California, Berkeley, CA, USA

A. A. Sulaimon, Department of Petroleum Engineering, Universiti Teknologi Petronas, Seri Iskandar, Perak, Malaysia

The SpringerBriefs series in Petroleum Geoscience & Engineering promotes and expedites the dissemination of substantive new research results, state-of-the-art subject reviews and tutorial overviews in the field of petroleum exploration, petroleum engineering and production technology. The subject focus is on upstream exploration and production, subsurface geoscience and engineering. These concise summaries (50–125 pages) will include cutting-edge research, analytical methods, advanced modelling techniques and practical applications. Coverage will extend to all theoretical and applied aspects of the field, including traditional drilling, shale-gas fracking, deepwater sedimentology, seismic exploration, pore-flow modelling and petroleum economics. Topics include but are not limited to:

- Petroleum Geology & Geophysics
- Exploration: Conventional and Unconventional
- Seismic Interpretation
- Formation Evaluation (well logging)
- Drilling and Completion
- Hydraulic Fracturing
- Geomechanics
- Reservoir Simulation and Modelling
- Flow in Porous Media: from nano- to field-scale
- Reservoir Engineering
- Production Engineering
- Well Engineering; Design, Decommissioning and Abandonment
- Petroleum Systems; Instrumentation and Control
- Flow Assurance, Mineral Scale & Hydrates
- Reservoir and Well Intervention
- Reservoir Stimulation
- Oilfield Chemistry
- Risk and Uncertainty
- Petroleum Economics and Energy Policy

Contributions to the series can be made by submitting a proposal to the responsible Springer contact, Charlotte Cross at charlotte.cross@springer.com or the Academic Series Editor, Prof Dorrik Stow at dorrik.stow@pet.hw.ac.uk.

More information about this series at http://www.springer.com/series/15391

Yongcun Feng · K. E. Gray

Lost Circulation and Wellbore Strengthening

 Springer

Yongcun Feng
Department of Petroleum and Geosystems
  Engineering
The University of Texas at Austin
Austin, TX
USA

K. E. Gray
Department of Petroleum and Geosystems
  Engineering
The University of Texas at Austin
Austin, TX
USA

ISSN 2509-3126   ISSN 2509-3134   (electronic)
SpringerBriefs in Petroleum Geoscience & Engineering
ISBN 978-3-319-89434-8   ISBN 978-3-319-89435-5   (eBook)
https://doi.org/10.1007/978-3-319-89435-5

Library of Congress Control Number: 2018939313

Printed on acid-free paper

This Springer imprint is published by the registered company Springer International Publishing AG
part of Springer Nature
The registered company address is: Gewerbestrasse 11, 6330 Cham, Switzerland

Preface

Lost circulation, the loss of partial or whole drilling fluid into the formation, is one of the most common and costly problems in drilling operations. Typical scenarios of lost circulation include drilling through pressure depleted zones, deepwater formations, naturally fractured shales, and carbonate formations. Wellbore strengthening is an effective and economic technique to prevent or mitigate lost circulation problem. This technique artificially increases the maximum pressure a wellbore can withstand by bridging or sealing the natural or drilling induced fractures on the wellbore wall. Although a number of experimental studies and field applications of wellbore strengthening have been reported in the drilling industry, the fundamental mechanisms of lost circulation and wellbore strengthening are still not thoroughly understood, and the industry still lacks sufficient models for lost circulation prediction and wellbore strengthening evaluation. This book makes an effort to fill these knowledge gaps.

This book focuses on the underlying mechanisms of lost circulation and wellbore strengthening. It presents a comprehensive, yet concise, overview of the fundamental studies on lost circulation and wellbore strengthening in the oil and gas industry, as well as a detailed discussion on the limitations of the wellbore strengthening methods currently used in the industry. The book provides several advanced analytical and numerical models for the simulations of lost circulation and wellbore strengthening under realistic conditions. Simulation results are presented to illustrate the capabilities of the models and to investigate the influences of key parameters. In addition, experimental results are also provided for better understanding of the subject.

The book is divided into six chapters. Chapter 1 begins with a brief introduction to the definition, scenarios and consequences of lost circulation, and the concept and different methods of wellbore strengthening. This chapter also provides a critical review of fundamental studies on lost circulation and wellbore strengthening.

Chapter 2 covers some background knowledges of drilling-related geomechanics which are closely related to lost circulation and wellbore strengthening. Concepts such as in situ stress, stress concentration around a wellbore, and drilling mud weight window are introduced. Familiarity with these concepts is essential for understanding the mechanisms of lost circulation and wellbore strengthening.

To understand fracture behavior during lost circulation, Chap. 3 describes a numerical model for lost circulation simulation. The model couples dynamic mud circulation in the wellbore and fracture propagation into the formation. It provides estimates of time-dependent wellbore pressure, fluid loss rate, and fracture profile during drilling. Numerical examples are presented to illustrate the effects of several key parameters on lost circulation.

Chapter 4 illustrates the mechanisms of wellbore strengthening in detail. An analytical model based on linear elastic fracture mechanics is introduced which provides a fast procedure to predict wellbore strengthening after bridging the fractures. Moreover, a numerical model is developed which gives a more detailed description of the distribution of local stress and fracture width with wellbore strengthening operations. A couple of experimental studies of wellbore strengthening are also described in this chapter.

Chapter 5 summarizes the properties and features of various lost circulation materials (LCMs) currently used in the drilling industry. Formations with different lithology usually require different LCMs. LCMs suitable for permeable sandstones, low-permeability shales, and carbonate formations are discussed in this chapter.

Chapter 6 provides some recommendations for future endeavors.

The book will help the readers understand quickly the concepts related to lost circulation and wellbore strengthening. It offers valuable information and guidelines for drilling engineers who face lost circulation problem in their wells and want to use wellbore strengthening technique to solve the problem. The book is also useful for industrial researchers and graduate students who perform fundamental researches in this area.

Austin, USA   Yongcun Feng
                                                                                                        K. E. Gray

Acknowledgements

The authors thank the Wider Windows Industrial Affiliate Program, the University of Texas at Austin, for financial and logistical support of the research work utilized in this book.

We gratefully acknowledge financial support from Wider Windows sponsors BHP Billiton, British Petroleum, Chevron, ConocoPhillips, Halliburton, Marathon, National Oilwell Varco, Occidental Oil and Gas, and Shell.

Yongcun Feng
K. E. Gray

Contents

designed, the pre-flush successfully removed the filter cake across a depleted sand, and total lost circulation was observed. A subsequent trip inside the casing to determine the fluid level confirmed that the hydrostatic pressure in the wellbore had stabilized at the estimated FPP in the depleted sand. It was also noted that the well had been drilled with an equivalent circulating density (ECD) significantly greater than this pressure, without significant fluid loss.

From these and multiple additional observations, it was determined that the tangential (hoop) stress equation (Kirsch solution) did a reasonably good job of estimating fracture initiation pressure (leak-off), including changes with respect to wellbore orientation and reservoir depletion, as long as the filter cake remained intact. However, with the filter cake removed, or absent, total losses occurred at FPP. Conversely, it was determined that for multiple shallower formations, which consisted mostly of silty shale and carbonates, leak-off tests were lower than expected, with seemingly no benefit from near wellbore hoop stress. It was surmised that the near wellbore stress field was bypassed by preexisting fractures, or there was a lack of filter cake development. It was also observed that there were often significant differences between FCP, FPP, and leak-off pressure, contrary to assertions in the literature.

It was suggested that the depleted formations were being strengthened, with fractures initiated at the wellbore wall and then quickly filled with background LCM, creating a plug at the fracture opening. But observed leak-off pressures were equal to (and not greater than) the hoop stress estimated by the Kirsch solution with no "strengthening." This discrepancy may be explained by an LCM enhanced filter cake forcing fractures to initiate at the wellbore wall, taking advantage of the already present hoop stress. There remains widespread disagreement about this, but interestingly, the LCM recommendations are identical in either case!

In the late 1990s and early 2000s, with a major exploration push into the deepwater Gulf of Mexico (GOM), operators immediately experienced difficulties associated with narrow drilling margins. The pore pressure/fracture gradient window had to be pushed to its limit in nearly every hole section, to ensure there would be an adequate number of casing and liner strings to reach the objective formations. Wellbore breathing (often called ballooning) and lost circulation were often observed at pressures significantly below the previous casing shoe test. Synthetic based mud (SBM) was very popular due to its shale inhibition, increased lubricity, reduced ECD, ease of running and maintaining, and significantly reduced propensity for differential sticking. However, it also had its drawbacks, including higher costs, cuttings disposal issues, potential masking of small gas influxes, and often lower observed fracture gradient compared to water-based mud (WBM). The widespread availability of logging while drilling (LWD) tools now made it possible to observe where wellbore breathing and lost circulation events were occurring. Interestingly, these losses were often seen in silty or ratty shale, often but not necessarily, at the transition between clean shale and sand. Furthermore, wellbore breathing was almost exclusively associated with nonaqueous drilling fluids. Several operators reported incidences of switching back to (WBM) in a particular hole section, in order to continue drilling with fewer mud losses and avoid setting casing prematurely. Significant work was being performed to describe the differences in FPP pressure between WBM and SBM, which focused on the major filter cake property differences within a fracture. The WBM filter cake was apparently superior for isolating the fracture tip from the wellbore pressure, thereby increasing FPP.

In addition, the numerous "lost circulation pills" formulated by various service companies for squeezing into and sealing fractures, required dewatering once in place in order for the fracture to close and seal around them. It was observed that these pills rarely performed well, unless they were mixed in a water-based carrier fluid. Otherwise, they would not effectively dewater into the shale and allow the fracture to close. This observation led to consideration of wettability of the silty shale formations (usually water wet) and associated capillary entry pressures for non-wetting phase SBM, as playing a significant role in lost circulation pill dewatering. It was also considered that this same phenomenon may be partially responsible for the improved filter cake properties for WBM compared to SBM, inside a fracture.

Despite improved understanding of wellbore breathing and lost circulation with SBM, results of mitigation strategies remain mixed. Pills that do not require dewatering are available, but operations to spot them are still expensive and time consuming, and these pills are frequently eroded by subsequent drilling operations. Some operators have focused more on ECD reduction methods for improving drilling margins, including constant bottom hole pressure techniques.

While lost circulation and its control have been studied extensively and significant advances have been achieved, knowledge gaps and unsolved problems still pose significant nonproductive time and costs for the drilling industry. Additional studies are needed, and topical recommendations by the authors of this monograph include: preexisting fractures; thermal effects; time-dependent developments of external and internal mudcake; numerical models to simulate transportation and deposition of LCMs in fractures; bridging/plugging processes; fracture geometry during drilling for selecting/adjusting size distribution of LCMs in real time; improved or new logging while drilling techniques for acquiring better knowledge of drilling-induced or preexisting natural fractures; lost circulation in carbonates; LC events in anisotropic/heterogeneous formations with complex lithology, stress, and pressure profiles; advanced LCMs that are both reservoir and environment friendly.

John Jones
Marathon Oil Company (retired)

Chapter 1
Introduction

Abstract Lost circulation is a wellbore instability problem that has plagued the drilling industry for years. Wellbore strengthening is recognized as an effective technique to prevent or mitigate lost circulation. This chapter introduces the common scenarios and consequences of lost circulation, as well as the concepts and different types of wellbore strengthening treatments. A review of fundamental studies on lost circulation and wellbore strengthening is also presented.

1.1 Lost Circulation

Lost circulation is the partial or whole losses of drilling fluid into the formation during drilling. It is a major cause of non-productive time (NPT) in the drilling industry. Lost circulation can lead to various drilling incidents, such as differential sticking and well control events, which further increase NPT and drilling costs [1]. More than 12% of NPT has been reported due to lost circulation in the Gulf of Mexico area shelf drilling [2]. The US Department of Energy reported that on average 10–20% of the drilling cost of high-pressure and high-temperature (HTHP) wells is expended on mud losses [3]. The impact of lost circulation on well construction is significant, representing an estimated 2–4 billion dollars annually in lost time, lost drilling fluid, and materials used to stem mud losses [4].

Mud loss may cause a reduced mud level in the well annulus. As a result, the bottom hole pressure (BHP) may become insufficient to balance formation pressure, and well control issues such as kick and underground blowouts will occur. Wellbore collapse may also occur due to reduced BHP. In some cases, the collapsed wellbore may result in buried drilling tools and stuck pipe [5, 6]. These incidents further increase NPT and drilling costs.

Most lost circulation events are due to fracture extension from the wellbore to the far field region. So lost circulation is a fracture initiation and propagation problem, occurring when the BHP is high enough to create fractures into the formation. Lost circulation commonly happens in formations with narrow drilling mud weight

© The Author(s) 2018
Y. Feng and K. E. Gray, *Lost Circulation and Wellbore Strengthening*,
SpringerBriefs in Petroleum Geoscience & Engineering,
https://doi.org/10.1007/978-3-319-89435-5_1

windows (MWW) between pore pressure/collapse pressure gradient and fracture gradient.

Several typical scenarios having a narrow MWW include depleted zones, deep-water formations, naturally fractured formations, and deviated wellbores [7]. In depleted zones, pore pressure reduction leads to a decreased fracture pressure and thus a lower pressure-bearing capacity of the wellbore. In deep-water formations, considerable water depth can cause a lower-than-usual fracture pressure. While drilling deviated wellbores, the MWW may diminish quickly with borehole inclination increase in some field stress conditions. Existence of natural fractures can significantly reduce the pressure-bearing capacity of a wellbore due to the low strength and reopening pressures of such fractures.

Note that lost circulation events are also commonly encountered in carbonate formations. This type of formation is often characterized by the presence of vugs and cavities [8–10]. Nevertheless, this book mainly focuses on lost circulation through fracture extension in clastic formations rather than through vugs in carbonate formations.

1.2 Wellbore Strengthening

The wellbore strengthening technique has been extensively used in the drilling industry to prevent or mitigate drilling fluid loss. Wellbore strengthening can be defined as methods to artificially increase the maximum pressure a wellbore can withstand without intolerable mud losses. Wellbore strengthening aims to enhance the effective fracture pressure and widen the mud weight window, rather than increasing the strength of the wellbore rock. By preventing and/or mitigating fluid loss, wellbore strengthening also reduces lost circulation associated NPT events, e.g. wellbore instability, pipe sticking, underground blowouts, and kicks.

Wellbore strengthening attempts to bridge, plug, or seal wellbore fractures with lost circulation materials (LCMs) to arrest the propagation of lost circulation fracture(s). The pressure-bearing capacity of the wellbore can be enhanced by one or a combination of the following mechanisms in wellbore strengthening treatments.

- Bridge a fracture near its mouth to increase the local compressive hoop stress around the wellbore and enhance fracture opening resistance [11–21];
- Widen and prop a fracture to enhance the fracture closure stress that acts on closing the fracture [22–29];
- Form a filter cake in the fracture to isolate the fracture tip from wellbore pressure and enhance resistance to fracture propagation [4, 30–39].

1.3 A Literature Review of Fundamental Studies on Lost Circulation and Wellbore Strengthening

1.3.1 Experimental Studies

Very few experimental studies have been conducted on lost circulation and wellbore strengthening. The DEA-13 experimental study conducted in the middle 1980s to early 1990s [33, 37, 40] is an early experimental investigation into lost circulation. The aim of that study was to examine why lost circulation occurs less frequently while drilling with water based mud (WBM) than with oil based mud (OBM). A major observation of DEA-13 project was that fracture propagation pressure (FPP) is strongly related to mud type and significantly increased by the use of LCM additives. This result was explained by a physical model called "tip screen-out" [33, 34, 36, 40], which indicates that the increase in FPP is due to isolation of the fracture tip and wellbore pressure by an LCM filtercake in the fracture.

The GPRI 2000 project conducted in the late 1990s to early 2000s [38] is another major experimental effort. The purpose of the GPRI 2000 project was to evaluate the capabilities of different LCMs on increasing fracture gradient. The experimental results show that fracture reopening pressure (FRP) of a wellbore can be increased by using LCMs and this effect is more remarkable in WBM than in OBM or synthetic based muds (SBM).

A recent experimental study on lost circulation conducted from late 2000s to early 2010s is called the Lost Circulation and Wellbore Strengthening Research Cooperative Agreement (RCA) project [41]. The aim of this project was to investigate the wellbore strengthening mechanism and the effectiveness of different wellbore strengthening methods (preventive and remedial methods). The main findings of this study include (1) a preventive wellbore strengthening treatment is more effective than remedial treatment; (2) particle size distribution (PSD) and concentration of LCM are critical in wellbore strengthening; and (3) fracture pressure achieved with wellbore strengthening can be higher than the formation breakdown pressure (FBP).

1.3.2 Physical Models of Wellbore Strengthening

There are three major physical models used in the drilling industry for explaining why wellbore strengthening treatments can "strengthen" a wellbore. They are the Stress Cage model, Fracture Closure Stress model, and Fracture Propagation Resistance model.

In the Stress Cage model [11], the LCMs wedge the fracture close to the wellbore together with filtration control agents. Next, trapped fluid in the fracture filters into the formation due to pressure difference; meanwhile compressive forces are transferred to the LCM-bridge at the fracture mouth. Finally, the fracture is bridged at the fracture mouth, resulting in an increased hoop stress.

In the Fracture Closure Stress (FCS) model [25], a fracture at the wellbore wall is first created and widened to increase the compressive stress (i.e. fracture closure stress) in the adjacent rock. Next, LCMs are forced into the fracture. As the carrier fluid leaks off into the formation, the LCM particles consolidate and form an immobile mass inside the fracture that keeps the fracture open and isolates the fracture tip from wellbore pressure. The increased fracture closure stress and isolation of the fracture tip make the fracture more difficult to open and extend.

The Fracture Propagation Resistance (FPR) model [34, 38] does not aim to increase wellbore hoop stress; instead it attempts to increase resistance against fracture propagation by forming a filtercake inside the fracture. The filtercake can seal the fracture tip and prevent pressure communication between the fracture tip and wellbore, thereby increasing the resistance to fracture propagation.

1.3.3   Analytical Studies

There is no extensive analytical investigation on lost circulation and wellbore strengthening in the drilling industry. Several parameters of interest in analytical studies are fracture pressure, fracture width, and fracture-tip stress intensity factor, before and after bridging/sealing the fracture. For calculating these parameters, it is important to consider the near wellbore stress concentration for short fractures and pressure drop along the fracture for large fractures.

[42] developed an approximate, closed-form model for fracture mouth opening of two fractures symmetrically located at the wellbore wall, based on linear elastic fracture mechanics. Their model is only valid for a relative short fracture with a length less than 4 wellbore radii and for a maximum to minimum horizontal stress ratio less than 2.

[1] proposed a semi-analytical solution for wellbore strengthening analysis based on singular integral formulation of the stress field. The model considers far field stress anisotropy and near wellbore stress perturbation, but it doesn't consider pressure drop along the fracture.

[43] applied the penny-shaped hydraulic fracture model developed by [44] to analyze fracture pressure increase after plugging the fracture. The model does not consider near wellbore stress concentration and therefore can only be used for a long fracture. Moreover, the model assumes that the pressure is uniform inside the fracture from the wellbore to the LCM plug (i.e. it does not consider pressure drop), which is not realistic for a large fracture from the perspective of viscous pressure drop along the fracture.

[36] presented two sets of closed-form solutions of stress intensity factor and fracture pressure for fractures bridging at the fracture mouth and inside the fracture, based on linear elastic fracture mechanics and the superposition principle. The model also assumes that the pressure inside the fracture from fracture mouth to bridge location is uniform and equal to wellbore pressure. The model for bridging inside the fracture neglects the effect of the wellbore on stress intensity factor induced by

pressure from the LCM bridge to fracture tip. Therefore, strictly speaking, it is only valid for the cases with large fractures when the effect of the wellbore can be ignored.

[39] extended the KGD hydraulic fracture model [45–47] to calculate fracture pressure and fracture width after sealing the fracture in wellbore strengthening. However, this model neglects both near wellbore stress concentration and pressure drop in the fracture.

1.3.4 Numerical Studies

Numerical methods have also been applied to simulate lost circulation and wellbore strengthening for understanding the mechanisms behind them.

[2, 20] developed a 2D boundary-element method (BEM) model to calculate stress and fracture width distribution before and after bridging a fracture in wellbore strengthening. [42] used a 2D finite-element method (FEM) model to investigate the fracture width distribution for two symmetrically located fractures on the wellbore with various in situ stresses and fracture lengths; but the model does not simulate fracture bridging. [11] employed a 2D FEM model to calculate fracture width and hoop stress distribution after bridging the fracture near its mouth under nearly isotropic in situ stresses. All of these numerical models assume that the rock is linearly elastic and do not consider porous features of the rock, therefore the effect of pore pressure and fluid flow are not considered.

In order to perform a comprehensive parametric study for wellbore strengthening, [16, 48] developed a 2D FEM poroelastic model considering fluid flow inside the rock and across fracture surfaces. That model is described later in this book. With the aim to exam the hypothesis of hoop stress increase when fractures are sealed/bridged as presented in the Stress Cage theory, [49, 50] modeled fracture propagation and sealing during lost circulation and wellbore strengthening using the cohesive zone model. They argued that fracture sealing/bridging cannot increase wellbore hoop stress beyond its ideal state when no fracture exists. In their model, a predefined injection rate (fluid loss rate) boundary condition was defined at the fracture inlet which is not consistent with the actual drilling situation where the downhole condition at the fracture inlet is neither a constant flow rate nor a constant pressure.

References

1. Shahri MP, Oar T, Safari R, Karimi M, Mutlu U (2014) Advanced geomechanical analysis of wellbore strengthening for depleted reservoir drilling applications. In: IADC/SPE Drilling Conference and Exhibition, Fort Worth, Texas, 4–6 March. SPE-167976-MS. http://dx.doi.org/10.2118/167976-MS
2. Wang H, Towler BF, Soliman MY (2007) Fractured wellbore stress analysis: sealing cracks to strengthen a wellbore. In: SPE/IADC Drilling Conference, 20–22 February, Amsterdam, The Netherlands. https://doi.org/10.2118/104947-MS

3. Growcock FB, Kaageson-Loe N, Friedheim J, Sanders MW, Bruton J (2009) Wellbore stability, stabilization and strengthening. In: Offshore Mediterranean conference and exhibition, 25–27 March, Ravenna, Italy

4. Cook J, Growcock F, Guo Q, Hodder M, Van Oort E (2011) Stabilizing the wellbore to prevent lost circulation. Oilfield Rev 23(4)

5. Lavrov A (2016) Lost circulation: mechanisms and solutions, 1st edn. Gulf Professional Publishing, Cambridge

6. Messenger J (1981) Lost circulation. Pennwell Corp, Tulsa

7. Feng Y, Jones JF, Gray KE (2016) A review on fracture-initiation and -propagation pressures for lost circulation and wellbore strengthening. SPE Drill Completion 31(02):134–144

8. Davidson E, Richardson L, Zoller S (2000) Control of lost circulation in fractured limestone reservoirs. In: IADC/SPE Asia Pacific Drilling Technology, 11–13 Sept, Kuala Lumpur, Malaysia. https://doi.org/10.2118/62734-MS

9. Masi S, Molaschi C, Zausa F, Michelez J (2011) Managing circulation losses in a harsh drilling environment: conventional solution vs. CHCD through a risk assessment. SPE Drilling Completion 26(02):198–207

10. Wang S et al (2010) Real-time downhole monitoring and logging reduced mud loss drastically for high-pressure gas wells in Tarim Basin, China. SPE Drilling Completion 25(02):187–192

11. Alberty MW, McLean MR (2004) A physical model for stress cages. In: SPE Annual Technical Conference and Exhibition, 26–29 Sept, Houston, Texas. https://doi.org/10.2118/90493-MS

12. Aston MS, Alberty MW, Duncum SD, Bruton JR, Friedheim JE, Sanders MW (2007) A new treatment for wellbore strengthening in shale. Presented at the SPE Annual Technical Conference and Exhibition, 11–14 Nov, Anaheim, California, USA. https://doi.org/10.2118/110713-MS

13. Aston MS, Alberty MW, McLean MR, de Jong HJ, Armagost K (2004) Drilling fluids for wellbore strengthening. Presented at the IADC/SPE Drilling Conference, 2–4 March, Dallas, Texas. https://doi.org/10.2118/87130-MS

14. Bassey A et al (2012) A new (3D MUDSYST Model) approach to wellbore strengthening while drilling in depleted sands: a critical application of LCM and stress caging model. In: Nigeria Annual International Conference and Exhibition, 6–8 August, Lagos, Nigeria. https://doi.org/10.2118/162946-MS

15. Chellappah K, Kumar A, Aston M (2015) Drilling depleted sands: challenges associated with wellbore strengthening fluids. In: SPE/IADC drilling conference and exhibition,17–19 March, London, England, UK. https://doi.org/10.2118/173073-MS

16. Feng Y, Arlanoglu C, Podnos E, Becker E, Gray KE (2015) Finite-element studies of hoop-stress enhancement for wellbore strengthening. SPE Drilling Completion 30(01):38–51

17. Loloi M, Zaki KS, Zhai Z, Abou-Sayed AS (2010) Borehole strengthening and injector plugging—the common geomechanics thread. In: North Africa technical conference and exhibition, 14–17 Feb, Cairo, Egypt. https://doi.org/10.2118/128589-MS

18. Shen X (2016) Widened safe mud window for naturally fractured shale formations: a workflow and case study. In: IADC/SPE Asia Pacific drilling technology conference, 22–24 August, Singapore. https://doi.org/10.2118/180519-MS

19. Song J, Rojas JC (2006) Preventing mud losses by wellbore strengthening. In: SPE Russian Oil and Gas Technical Conference and Exhibition, 3–6 October, Moscow, Russia. https://doi.org/10.2118/101593-MS

20. Wang H, Soliman MY, Towler BF (2008) Investigation of factors for strengthening a wellbore by propping fractures. In: IADC/SPE drilling conference, 4–6 March, Orlando, Florida, USA. https://doi.org/10.2118/112629-MS

21. Zhang J, Alberty M, Blangy JP (2016) A semi-analytical solution for estimating the fracture width in wellbore strengthening applications. In: SPE deepwater drilling and completions conference, 14–15 Sept, Galveston, Texas, USA. https://doi.org/10.2118/180296-MS

22. Buechler S et al (2015) Root cause analysis of drilling lost returns in injectite reservoirs. In: SPE annual technical conference and exhibition, 28–30 Sept, Houston, Texas, USA. https://doi.org/10.2118/174848-MS

23. Duffadar RD, Dupriest FE, Zeilinger SC (2013) Practical guide to lost returns treatment selection based on a holistic model of the state of the near wellbore stresses. In: SPE/IADC drilling conference, 5–7 March, Amsterdam, The Netherlands. https://doi.org/10.2118/163481-MS

24. Dupriest FE (2009) Use of new hydrostatic packer concept to manage lost returns, well control, and cement placement in field operations. SPE Drilling Completion 24(4):574–580

25. Dupriest FE (2005) Fracture closure stress (FCS) and lost returns practices. In: SPE/IADC Drilling Conference, 23–25 Feb, Amsterdam, Netherlands. https://doi.org/10.2118/92192-MS

26. Dupriest FE, Smith MV, Zeilinger SC, Shoykhet N (2008) Method to eliminate lost returns and build integrity continuously with high-filtration-rate fluid. In: IADC/SPE drilling conference, 4–6 March, Orlando, Florida, USA. https://doi.org/10.2118/112656-MS

27. Lai YK, Woodward B (2014) Fracture closure stress treatment in low fracture pressure reservoir. In: International petroleum technology conference, 10–12 Dec, Kuala Lumpur, Malaysia. https://doi.org/10.2523/IPTC-17850-MS

28. Montgomery JK, Keller SR, Krahel N, Smith MV (2007) Improved method for use of chelation to free stuck pipe and enhance treatment of lost returns. In: SPE/IADC drilling conference, 20–22 Feb, Amsterdam, The Netherlands. https://doi.org/10.2118/105567-MS

29. Vaczi K, Estes BL, Ryan J, Linehan S, Mota M (2009) Drilling fluid design prevents lost returns by building integrity continuously while drilling in East Texas. In: SPE/IADC drilling conference and exhibition, 17–19 March, Amsterdam, The Netherlands. https://doi.org/10.2118/119269-MS

30. Contreras O, Hareland G, Husein M, Nygaard R, Alsaba M (2014) Wellbore strengthening in sandstones by means of nanoparticle-based drilling fluids. In: SPE deepwater drilling and completions conference, 10–11 Sept, Galveston, Texas, USA. https://doi.org/10.2118/170263-MS

31. Fuh GF, Beardmore DH, Morita N (2007) Further development, field testing, and application of the wellbore strengthening technique for drilling operations. In: SPE/IADC drilling conference, 20–22 Feb, Amsterdam, The Netherlands. https://doi.org/10.2118/105809-MS

32. Jacobs T (2014) Pushing the frontier through wellbore strengthening. J Petrol Technol 66(11):64–73

33. Morita N, Black AD, Fuh GF (1996) Borehole breakdown pressure with drilling fluids—I. Empirical results. Int J Rock Mech Min Sci Geomech Abstr 33(1):39–51

34. Morita N, Fuh GF, Black AD (1996) Borehole breakdown pressure with drilling fluids—II. Semi-analytical solution to predict borehole breakdown pressure. Int J Rock Mech Min Sci Geomech Abstr 33(1):53–69

35. Morita N, Black AD, Guh GF (1990) Theory of lost circulation pressure. In: SPE annual technical conference and exhibition, 23–26 Sept, New Orleans, Louisiana. https://doi.org/10.2118/20409-MS

36. Morita N, Fuh G-F (2012) Parametric analysis of wellbore-strengthening methods from basic rock mechanics. SPE Drilling Completion 27(02):315–327

37. Onyia EC (1994) Experimental data analysis of lost-circulation problems during drilling with oil-based mud. SPE Drilling Completion 9(01):25–31

38. van Oort E, Friedheim JE, Pierce T, Lee J (2011) Avoiding losses in depleted and weak zones by constantly strengthening wellbores. SPE Drilling Completion 26(4):519–530

39. van Oort E, Razavi OS (2014) Wellbore strengthening and casing smear: the common underlying mechanism. In: IADC/SPE Drilling Conference and Exhibition, 4–6 March, Fort Worth, Texas, USA. https://doi.org/10.2118/168041-MS

40. Fuh GF, Morita N, Boyd PA, McGoffin SJ (1992) A new approach to preventing lost circulation while drilling. In: SPE Annual Technical Conference and Exhibition, 4–7 Oct, Washington, D.C. https://doi.org/10.2118/24599-MS

41. Guo Q, Cook J, Way P, Ji L, Friedheim JE (2014) A comprehensive experimental study on wellbore strengthening. In: IADC/SPE Drilling Conference and Exhibition, 4–6 March, Fort Worth, Texas, USA. https://doi.org/10.2118/167957-MS

42. Guo Q, Feng YZ, Jin ZH (2011) Fracture aperture for wellbore strengthening applications. In: 45th U.S. rock mechanics/geomechanics symposium, 26–29 June, San Francisco, California.

43. Ito T, Zoback MD, Peska P (2001) Utilization of mud weights in excess of the least principal stress to stabilize wellbores: theory and practical examples. SPE Drilling Completion 16(4):221–229
44. Abé H, Keer LM, Mura T (1976) Growth rate of a penny-shaped crack in hydraulic fracturing of rocks, 2. J Geophys Res 81(35):6292–6298
45. Geertsma J, De Klerk F (1969) A rapid method of predicting width and extent of hydraulically induced fractures. J Petrol Technol 21(12):1571–1581
46. Linkov AM (2013) Analytical solution of hydraulic fracture problem for a non-Newtonian fluid. J Min Sci 49(1):8–18
47. Christianovich SA, Zheltov YP (1955) Formation of vertical fractures by means of a highly viscous liquid. In: Proc. 4th World Petroleum Congress 2:579–586
48. Feng Y, Gray KE (2016) A parametric study for wellbore strengthening. J Nat Gas Sci Eng 30:350–363
49. Salehi S (2012) Numerical simulations of fracture propagation and sealing: implications for wellbore strengthening. Doctoral Dissertations, Missouri University of Science and Technology, Rolla, Missouri, USA
50. Salehi S, Nygaard R (2011) Numerical study of fracture initiation, propagation, sealing to enhance wellbore fracture gradient. In: 45th U.S. rock mechanics/geomechanics symposium, 26–29 June, San Francisco, California

Chapter 2
Drilling Related Geomechanics

Abstract Analyses of lost circulation and wellbore strengthening require knowledge of formation and wellbore geomechanics. This chapter provides a brief overview of some geomechanical concepts related to the subject, including in-situ stress, pore pressure, wellbore stress concertation, wellbore failure criteria, and drilling mud weight window. A variety of field injectivity tests for determining in-situ stress and fracture properties (e.g. fracture initiation, propagation, and closure pressures) are also discussed.

2.1 In-Situ Stress and Pore Pressure

In-situ stresses in a formation are usually denoted by three orthogonal principal stresses, which are the vertical principal stress S_v, the maximum horizontal stress S_{Hmax}, and the minimum horizontal stress S_{hmin}.

S_v is equal to the integration of rock density with depth [1]:

$$S_v = \int_0^z \rho(z)gdz \approx \bar{\rho}gz \tag{2.1}$$

where $\rho(z)$ is the density of rock, g is the gravitational acceleration, $\bar{\rho}$ is the average overburden rock density, z is depth.

In geologically relaxed areas, the horizontal stresses are nearly isotropic and can be expressed as a function of vertical stress and Poisson's ratio as [2]:

$$S_{Hmax} = S_{hmin} = \frac{v}{1-v}S_v \tag{2.2}$$

where v is the Poisson's ratio.

© The Author(s) 2018
Y. Feng and K. E. Gray, *Lost Circulation and Wellbore Strengthening*,
SpringerBriefs in Petroleum Geoscience & Engineering,
https://doi.org/10.1007/978-3-319-89435-5_2

For anisotropic conditions, considering pore pressure and thermal effect, the following equations can be used to determine horizontal principal stresses:

$$S_{xx} = \frac{v}{1-v}S_v + \frac{1-2v}{1-v}\alpha_p p_p + \frac{E\varepsilon_{xx}}{1-v^2} + \frac{E\alpha_T \Delta T}{1-v}$$

$$S_{yy} = \frac{v}{1-v}S_v + \frac{1-2v}{1-v}\alpha_p p_p + \frac{vE\varepsilon_{xx}}{1-v^2} + \frac{E\alpha_T \Delta T}{1-v} \qquad (2.3)$$

If $\varepsilon_{xx} < 0$, $S_{hmin} = S_{xx}$ and $S_{Hmax} = S_{yy}$

If $\varepsilon_{xx} > 0$, $S_{hmin} = S_{yy}$ and $S_{Hmax} = S_{xx}$

where α_p is the Biot coefficient, p_p is pore pressure, E is Young's modulus, ε_{xx} is tectonic strain, α_T is thermal expansion coefficient, ΔT is temperature difference between the depth of interest and surface.

The in-situ stress orientation can be estimated by one or a combination of the methods of: leak-off tests, stress-induced wellbore breakouts, stress-induced tensile wellbore fractures, hydraulic fracture orientations, earthquake focal plane mechanisms, and shear velocity anisotropy. Interested readers can refer to [1] for a detailed review of these methods.

In a normal pressure formation, the pore pressure is equal to the hydrostatic pressure of formation fluids and is expressed as [1]:

$$p_p = \int_0^z \rho_f(z)\, g\, dz \approx \rho_f g z \qquad (2.4)$$

where $\rho_f(z)$ is the density of formation fluids, ρ_f is the average fluid density.

Normal pressure systems develop when the rate of formation deposition does not exceed the rate of fluid escape from the pores. If the formation deposition rate is faster than fluid escape rate, abnormal pressure develops in the formation. On the other hand, subnormal formation pressure is also observed in oil and gas reservoirs which is usually caused by production operations.

2.2 Field Injectivity Tests

Precise determinations of in-situ stress and fracture parameters (e.g. fracture initiation and propagation pressures) are critical for lost circulation prevention and some other aspects, such as wellbore stability evaluation and casing program design [3, 4]. Field injectivity tests are usually conducted during drilling operations to assess in-situ stress, including formation integrity test (FIT), leak-off test (LOT), extended leak-off test (XLOT), and pump-in and flow-back test (PIFB) [5]. However, not all these tests can provide reliable stress information, either due to insufficient injection time/volume or due to a number of factors distorting test signatures and leading to interpretation difficulties and uncertainties [6]. A brief description of the different kinds of field injectivity tests is provided as follows.

- **Formation integrity test (FIT)** is to confirm the cement integrity near the casing shoe, and to ensure the formation strength at the shoe is sufficient to withstand any expected or potential loads while drilling the subsequent hole section. In an FIT, the wellbore pressure is increased to a pre-set value lower than fracture initiation pressure (FIP). Fracture pressure and field stress cannot be obtained because no fracture has been created.
- **Leak-off test (LOT)** measures the fracture initiation pressure (FIP) near the casing shoe. In an LOT, wellbore pressure is increased until a noticeable deviation from linearity appears on the pressure-time curve. The wellbore pressure at the deviation point is defined as leak-off pressure (LOP), which is commonly assumed equal to FIP and used to determine the upper limit of the drilling mud weight window. LOTs rarely provide far-field stress information because the fractures created in the tests remain very short and are dominated by the near-wellbore stress concentration.
- **Extended leak-off test (XLOT)** creates long enough fracture to measure far-field minimum horizontal stress S_{hmin}. In an XLOT, pumping continues until a relatively steady fracture propagation pressure (FPP) is reached, followed by a shut-in phase. Fracture closure pressure (FCP) can be predicted using the shut-in data and is commonly taken as the best estimate of S_{hmin}. However, XLOTs are usually conducted in "impermeable" tight shale with "dirty" drilling mud, where the fracture may not close due to limited leak-off, leading to difficulties in FCP prediction.
- **Pump-in and flow-back test (PIFB)** provides better prediction of FCP and S_{hmin}. A PIFB test is an XLOT followed by a flow-back phase, usually with a constant flow-back rate. FCP and thus S_{hmin} can be readily and accurately determined using the flow-back data, because fracture closure can be assured with fluid flow-back and is not dependent on fluid leak-off anymore. Therefore, PIFB tests provide a superior method of measuring S_{hmin} in impermeable formations. Unfortunately, PIFB tests are rarely performed in drilling operations, due to additional time and cost required to carry out the tests and the fear of damaging wellbore integrity [7].

A schematic wellbore pressure versus time plot of a field injectivity test with three stages (pump-in, shut-in, and flow-back) is shown in Fig. 2.1. During pump-in, wellbore pressure increases linearly to LOP, after which the curve deviates to the right. Beyond LOP, the pressure increases with a lower rate to formation breakdown pressure (FBP). After FBP, the pressure experiences a sudden drop and finally stabilizes at a relatively constant FPP. Following shut-in, the pressure first has another sudden drop from FPP to instantaneous shut-in pressure (ISIP), and then declines at a rate dependent on the permeability of the formation. In permeable formations, the facture may close with leak-off and a FCP can be determined as shown in Fig. 2.1. However, in impermeable formations, such a FCP cannot be obtained because the leak-off is too slow for the fracture to close within a reasonable amount of time. In the flow-back phase, the pressure first decreases with a smaller rate before FCP, and then with a larger rate after FCP.

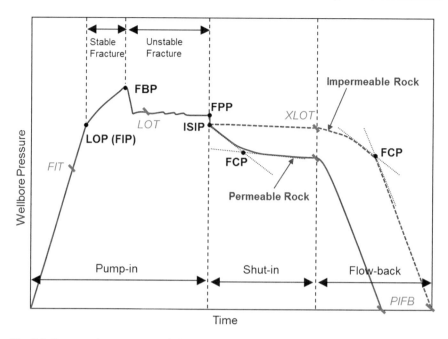

Fig. 2.1 Pressure-time response of field injectivity test (FIT—formation integrity test; LOT—leak-off test; XLOT—extended leak-off test; PIFB—pump-in and flow-back test; LOP—leak-off pressure; FIP—fracture initiation pressure; FBP—formation breakdown pressure; FPP—fracture propagation pressure; ISIP—instantaneous shut-in pressure; FCP—fracture closure pressure) (after [5], with permission from Springer)

A schematic wellbore pressure versus volume plot of a field injectivity tests in permeable and impermeable formations is shown Fig. 2.2. The horizontal axis is the volume of injection fluid added into the system during the test. The featured points, such as LOP, FPP and FCP, indicated in Fig. 2.2 correspond to those in Fig. 2.1.

2.3 Stress Around a Wellbore

For wellbore failure analysis, the stresses around a wellbore should be determined first. This is usually accomplished by transforming the in-situ stresses from the geodetic coordinate system (X, Y, Z) (Fig. 2.3a) to the wellbore Cartesian coordinate system (X′, Y′, Z′) (Fig. 2.3b), and then to the wellbore cylindrical coordinate system (r, θ, z) (Fig. 2.3c). α and i are the azimuth and inclination of the wellbore, respectively. The transformations of stresses are as follows.

The in-situ far-field stress tensor in coordinate (X, Y, Z) is:

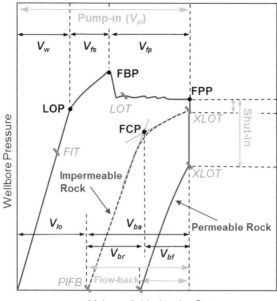

Volume Added to the System

Fig. 2.2 Pressure-volume response of field injectivity test (Vpi—total volume pumped into the system during the pump-in stage; Vw—injected volume before the fracture is initiated; Vfs—fluid volume added to the system during stable fracture propagation equal to volume flowing into the fracture plus any further volume change due to mud compression, casing expansion, fluid penetration and open hole expansion with wellbore pressure increase; Vfp—equal to the volume to extend the fracture and the volume loss into the formation due to leak-off during unstable fracture propagation; Vba—total flow-back volume; Vbf—flow back volume from the fracture; Vbr—returned volume due to inward fluid flow from the formation to the wellbore and wellbore shrinkage with pressure decrease; Vlo—total fluid volume lost into the formation) (after [5], with permission from Springer)

$$S = \begin{bmatrix} S_{Hmax} & 0 & 0 \\ 0 & S_{hmin} & 0 \\ 0 & 0 & S_v \end{bmatrix} \tag{2.5}$$

The stress in the wellbore Cartesian coordinate can be expressed as:

$$S' = LSL^T = \begin{bmatrix} S'_{xx} & S'_{xy} & S'_{xz} \\ S'_{yx} & S'_{yy} & S'_{yz} \\ S'_{zx} & S'_{zy} & S'_{zz} \end{bmatrix} \tag{2.6}$$

where,

Fig. 2.3 Coordinate
transformation. **a** geodetic
coordinate system (X, Y, Z);
b wellbore Cartesian
coordinate system
$(X^{'}, Y^{'}, Z^{'})$; **c** wellbore
cylindrical coordinate
system (r, θ, z)

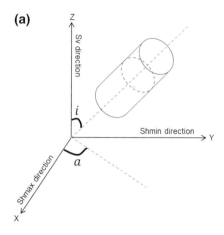

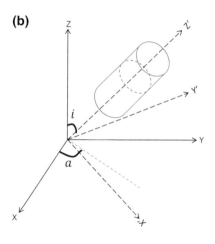

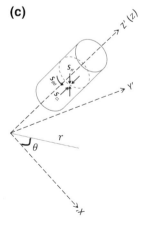

$$L = \begin{bmatrix} \cos(\alpha)\cos(i) & \sin(\alpha)\cos(i) & -\sin(i) \\ -\sin(\alpha) & \cos(\alpha) & 0 \\ \cos(\alpha)\sin(\alpha) & \sin(\alpha)\sin(i) & \cos(i) \end{bmatrix} \quad (2.7)$$

The stress around the wellbore in the wellbore cylindrical coordinate can be obtained as:

$$S_{rr} = \frac{S'_{xx} + S'_{yy}}{2}\left(1 - \frac{R^2}{r^2}\right) + \frac{S'_{xx} - S'_{yy}}{2}\left(1 + 3\frac{R^4}{r^4} - 4\frac{R^2}{r^2}\right)\cos(2\theta)$$
$$+ S'_{xy}\left(1 + 3\frac{R^4}{r^4} - 4\frac{R^2}{r^2}\right)\sin(2\theta) + p_w\frac{R^2}{r^2} \quad (2.8)$$

$$S_{\theta\theta} = \frac{S'_{xx} + S'_{yy}}{2}\left(1 + \frac{R^2}{r^2}\right) - \frac{S'_{xx} - S'_{yy}}{2}\left(1 + 3\frac{R^4}{r^4}\right)\cos(2\theta)$$
$$- S'_{xy}\left(1 + 3\frac{R^4}{r^4}\right)\sin(2\theta) - p_w\frac{R^2}{r^2} \quad (2.9)$$

$$S_{zz} = S'_{zz} - v\left[2\left(S'_{xx} - S'_{yy}\right)\frac{R^2}{r^2}\cos(2\theta) + 4S'_{xy}\frac{R^2}{r^2}\sin(2\theta)\right] \quad (2.10)$$

$$S_{r\theta} = \frac{S'_{yy} - S'_{xx}}{2}\left(1 - 3\frac{R^4}{r^4} + 2\frac{R^2}{r^2}\right)\sin(2\theta) + S'_{xy}\left(1 - 3\frac{R^4}{r^4} + 2\frac{R^2}{r^2}\right)\cos(2\theta) \quad (2.11)$$

$$S_{\theta z} = \left(-S'_{xz}\sin(\theta) + S'_{yz}\cos(\theta)\right)\left(1 + \frac{R^2}{r^2}\right) \quad (2.12)$$

$$S_{rz} = \left(S'_{xz}\cos(\theta) + S'_{yz}\sin(\theta)\right)\left(1 - \frac{R^2}{r^2}\right) \quad (2.13)$$

where R is the wellbore radius.

The stress components on the wellbore wall ($r = R$) can be written as:

$$S_{rr} = p_w \quad (2.14)$$

$$S_{\theta\theta} = S'_{xx} + S'_{yy} - 2\left(S'_{xx} - S'_{yy}\right)\cos(2\theta) - 4S'_{xy}\sin(2\theta) - p_w \quad (2.15)$$

$$S_{zz} = S'_{zz} - v\left[2\left(S'_{xx} - S'_{yy}\right)\cos(2\theta) + 4S'_{xy}\sin(2\theta)\right] \quad (2.16)$$

$$S_{r\theta} = 0 \quad (2.17)$$

$$S_{\theta z} = 2\left(-S'_{xz}\sin(\theta) + S'_{yz}\cos(\theta)\right) \quad (2.18)$$

$$S_{rz} = 0 \quad (2.19)$$

The radial stress S_{rr} is one of the principal stresses on the wellbore wall. The other two principal stresses can be determined as:

$$S_{prin1,2} = \frac{(S_{\theta\theta} + S_{zz})}{2} \pm \sqrt{\left(\frac{S_{\theta\theta} - S_{zz}}{2}\right)^2 + S_{\theta z}^2} \tag{2.20}$$

2.4   Wellbore Failure Criteria

The critical wellbore pressure required to create a fracture on the wellbore is defined as the fracture pressure of the wellbore. It can be determined by the tensile failure criterion:

$$S_3 - \alpha \cdot P_p \le S_t \tag{2.21}$$

where S_3 is the minimum principal stress on the wellbore wall; S_t is the tensile strength of the rock, which is a negative value.

The fracture pressure of an impermeable wellbore is different from that of a permeable wellbore because of fluid penetration through wellbore wall in the latter case. The fracture pressure of an impermeable vertical wellbore is given by [8]:

$$p_f = 3S_{hmin} - S_{Hmax} - p_p + S_t \tag{2.22}$$

The fracture pressure of a permeable vertical wellbore is given by [9]:

$$p_f = \frac{3S_{hmin} - S_{Hmax} - \eta p_p + S_t}{2 - \eta} \tag{2.23}$$

where $\eta = \alpha_p \left(\frac{1-2v}{1-v}\right)$ is a poroelastic parameter of the rock, which determines the magnitude of the stress induced by fluid penetration, and varies in the range [0, 1], from zero fluid penetration to unimpeded fluid penetration, respectively; α_p is Biot's coefficient; and v is Poisson's ratio.

Pre-existing cracks on the wellbore wall can significantly change the fracture pressure. [10] presented the following solution of fracture pressure of a wellbore with a pre-existing crack. This solution is obtained based on the Barenblatt condition, which dictates a balance between the tensile stress intensity factor produced by fluid pressure in the fracture and the negative stress intensity factor caused by the compressive in-situ stress:

$$p_f = \frac{3S_{hmin} - S_{Hmax}}{2} + \frac{K_{Ic}}{\pi\sqrt{2L}} \tag{2.24}$$

where K_{IC} and L are the fracture toughness and the length of the existing fracture, respectively. A limitation of this solution is that it is not suitable for very short fracture length.

Wellbore instability occurs when the wellbore pressure is too low to support the wellbore wall and shear wellbore failure occurs. The critical state for wellbore instability can be written as follows based on the Mohr-Coulomb theory:

$$S_1 - \alpha p_p = \left(S_3 - \alpha p_p\right) \tan^2 \left(\frac{\pi}{4} + \frac{\phi}{2}\right) + S_c \qquad (2.25)$$

where S_1 is the maximum principal stress on the wellbore wall; S_c is the uniaxial compressive strength of the rock; ϕ is the angle of internal friction.

2.5 Drilling Mud Weight Window

Drilling mud weight window is defined as the margin between the maximum mud weight before the occurrence of lost circulation and the minimum mud weight to balance formation pore pressures or avoid excessive wellbore failure. Lost circulation events commonly occur in formations with narrow drilling mud weight window.

As mentioned previously, two typical scenarios of lost circulation are fluid losses in depleted reservoirs and deep-water formations. This is because in depleted sands the reduction in pore pressure results in a corresponding reduction in fracture gradient. Conversely, the bounding and inter-bedded shale layers, as well as any isolated and un-drained sands, will maintain their original pore pressure and fracture gradient [11]. Therefore, as shown in Fig. 2.4, it may be difficult or impossible to reduce the drilling fluid density sufficiently to maintain equivalent circulating densities (ECD) below the depleted zone fracture gradient.

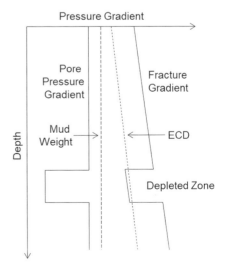

Fig. 2.4 Pore pressure and fracture gradient plot in depleted zone. Pore pressure decrease leads to a decrease in fracture gradient (after [11], with permission from SPE)

Fig. 2.5 Pore pressure and
fracture gradient plot in
deep-water formation with
abnormally high pressure.
There is a reduced
mud-weight window (after
[11], with permission from
SPE)

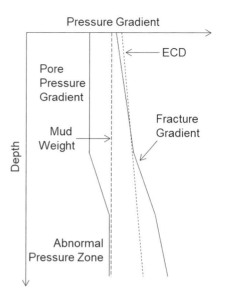

In deep-water formations, the total vertical stress is relatively low since sea water
does not provide as much overburden loading as sediment and rock. A reduction
in total vertical stress also results in a lower lateral stress and fracture gradient. If
abnormal pressures are also present, the mud-weight window may be very narrow,
as shown in Fig. 2.5. Under these circumstances, it may be challenging to avoid
hydraulic fracturing both while tripping due to surge/swab effects, and while circulating due to high annular friction losses and ECDs.

References

1. Zoback MD (2010) Reservoir geomechanics. Cambridge University Press
2. Fjar E, Holt RM, Raaen AM, Risnes R, Horsrud P (2008) Petroleum related rock mechanics.
   Elsevier
3. Feng Y, Gray KE (2017) Parameters controlling pressure and fracture behaviors in field injectivity tests: a numerical investigation using coupled flow and geomechanics model. Comput
   Geotech 87:49–61
4. Feng Y, Gray KE (2016) A comparison study of extended leak-off tests in permeable and
   impermeable formations. In: 50th U.S. rock mechanics/geomechanics symposium
5. Feng Y, Gray KE (2017) Discussion on field injectivity tests during drilling. Rock Mech Rock
   Eng 50(2):493–498
6. Feng Y, Jones J, Gray K (2015) Pump-in and flow-back tests for determination of fracture
   parameters and in-situ stresses. In: 2015 AADE national technical conference and exhibition,
   San Antonio
7. Ziegler F, Jones J (2014) Predrill pore-pressure prediction and pore pressure and fluid loss monitoring during drilling: a case study for a deepwater subsalt Gulf of Mexico well and discussion
   on fracture gradient, fluid losses, and wellbore breathing. Interpretation 2(1):SB45–SB55

8. Hubbert MK, Willis DG (1972) Mechanics of hydraulic fracturing. Petroleum Transactions, AIME, 210:153–168
9. Haimson B, Fairhurst C (1969) Hydraulic fracturing in porous-permeable materials. J Petrol Technol 21(07):811–817
10. Lee D, Bratton T, Birchwood R (2004) Leak-off test interpretation and modeling with application to geomechanics. In: Gulf Rocks 2004, the 6th North America Rock Mechanics Symposium (NARMS)
11. Feng Y, Jones JF, Gray KE (2016) A review on fracture-initiation and-propagation pressures for lost circulation and wellbore strengthening. SPE Drilling Completion 31(02):134–144

Chapter 3
Lost Circulation Models

Abstract The ability to model fracture growth during lost circulation is critical for loss prevention and wellbore strengthening design. Such a model can provide useful information for the optimization of drilling fluid rheology, well configuration, pump schedule, and LCM particle size distribution. In this chapter, a lost circulation simulation model is developed, based on the finite-element method. The model successfully couples mud circulation in the wellbore, fracture propagation, fracture fluid flow, and the surrounding reservoir behavior during a lost circulation event. It can predict time-dependent fluid loss rate, wellbore pressure, and fracture geometry. Borehole ballooning is a phenomenon closely related to lost circulation and is often regarded as an omen of a lost circulation event during drilling. Using a similar numerical approach, a model for wellbore ballooning in naturally fractured formations is also proposed in this chapter. The model is able to capture dynamic fracture growth during mud circulation when downhole circulation pressure is higher than fracture reopening pressure of the natural fractures, as well as fracture closure during the pump-off period when downhole pressure is lower. Time-dependent wellbore pressure, fluid loss/gain rate, and fracture profile in borehole ballooning can be obtained. The model aids in understanding the mechanisms involved in wellbore ballooning in naturally fractured formations.

3.1 Introduction

Mud losses through natural or drilling-induced fractures constitute the majority of lost circulation events during drilling, especially for drilling operations in depleted reservoirs, deep-water wells, and fractured shales [1, 2]. Knowledge of fluid loss and associated fracture growth behavior can help understand how to prevent this problem [3]. In addition, accurate prediction of fracture geometry is important for the selection of lost circulation materials (LCMs) to plug or bridge the fractures to mitigate fluid losses [4].

© The Author(s) 2018

Y. Feng and K. E. Gray, *Lost Circulation and Wellbore Strengthening*,
SpringerBriefs in Petroleum Geoscience & Engineering,
https://doi.org/10.1007/978-3-319-89435-5_3

A few models have been developed for fluid losses into natural fractures. The existing models usually assume a pre-defined fracture with a fixed length (i.e. fracture is not growing) and an impermeable fracture surface (i.e. leak-off through the fracture surface is ignored) [5–11]. Usually, fluid flow within the fracture is modeled based on Reynolds lubrication theory, and the normal deformation of fracture surfaces (i.e., fracture width change) is described by a linear or exponential deformation model which relates fracture width and fluid pressure in the fracture [10]. Even with the significant simplifications incorporated in these models, they have proven to be very useful as diagnostic tools to estimate the width of the natural fracture for design of LCMs [12].

For mud losses into drilling-induced fractures, [13] and [4] modeled the fracture growth using the cohesive zone fracture model implemented in a finite element code. However, in their models, a constant or user-specified time-dependent injection rate boundary condition at the bottom of the hole were used to drive the fractures. While such an injection rate driven fracture can capture the physics of an injectivity test or a hydraulic fracturing treatment, it cannot capture the physics of a drilling-induced fracture [4]. The condition at the fracture mouth of a drilling-induced fracture is a dynamic bottom hole pressure (BHP) or ECD while drilling, rather than a constant flow rate. These two boundary conditions can result in significantly different fracture geometries [4]. Moreover, there is no way to capture the amount of lost and returned circulation with an injection rate boundary condition because all the injected fluid is forced into the fracture.

A comprehensive lost circulation model should couple mud circulation in the wellbore for instantaneous prediction of BHP, rather than define a pressure boundary condition at the hole bottom. In so doing, the dynamic BHP, fluid loss, and time-dependent fracture geometry can be quantified. This chapter illustrates a numerical framework that allows for coupling between mud circulation in the wellbore and induced fracture propagation into the formation. The new model, therefore, provides a unique method for solving the challenging bottom-hole boundary conditions in modeling drilling-induced fractures.

The coupled lost circulation model is developed by combining a cohesive zone model (CZM) for simulating fracture propagation and a pipe flow model for simulating mud circulation through the drill pipe and annulus. The model is implemented with the finite-element software package Abaqus [14]. A similar model is also developed to simulate borehole ballooning events, phenomena closely related to lost circulation.

3.2 Modeling Study of Lost Circulation

As shown in Fig. 3.1, a lost circulation system generally consists of three major components: the well, the fracture, and the formation. The following physical processes occurring in a lost circulation event are included and simulated simultaneously in the proposed lost circulation model [3]:

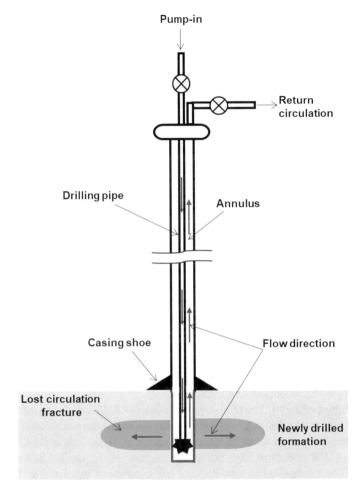

Fig. 3.1 Illustration of lost circulation system with well, formation and fracture

- Mud circulation in the well,
- Lost circulation fracture propagation and fluid flow in the fracture,
- Formation rock deformation and pore fluid flow.

The lost circulation model is developed on the finite-element platform Abaqus, which is a general purpose finite-element code for solving linear and non-linear stress-analysis problems [14].

3.2.1 Theory

Mud circulation in the well. Mud circulation in the wellbore is modeled based on Bernoulli's equation (using a Darcy-Weisbach approach) considering both gravity and viscous pressure losses [14]:

$$\Delta P - \rho g \Delta Z = \frac{f L}{D_h} \frac{\rho v^2}{2} \tag{3.1}$$

where ΔP is the pressure difference; ΔZ is the elevation difference; v is the fluid velocity in the pipe; ρ is the fluid density; g is the gravity acceleration factor; L is the pipe length; f is the friction factor; $D_h = \frac{4A}{S}$ is the hydraulic diameter of the pipe; A is the cross-sectional area of the pipe; and S is the wetted perimeter of the pipe.

The friction factor f in Eq. (3.1) is an important parameter which determines the friction loss during fluid flow. In this work, the friction factor is determined by the Churchill friction loss formula [15]:

$$f = 8 \left[\left(\frac{8}{Re} \right)^{12} + \frac{1}{(A + B)^{1.5}} \right]^{\frac{1}{12}} \tag{3.2}$$

where $A = \left[-2.457 ln \left(\left(\frac{7}{Re} \right)^{0.9} + 0.27 \frac{K_s}{D_h} \right) \right]^{16}$; $B = \left(\frac{37350}{Re} \right)^{16}$; Re is the Reynolds number; K_s is the roughness of the pipe.

The Fracture Model. The cohesive zone model (CZM) is used to model fracture growth. In contrast to the bulk-continuum materials for which constitutive behaviors are usually described in terms of stress and strain, the constitutive behavior of the fracture in CZM is defined in terms of the traction and separation of the fracture interface. The traction-separation damage law consists of three stages, i.e. initial loading before damage, damage initiation, and damage evolution.

Figure 3.2 shows a bilinear traction-separation law used in this study. The initial loading process before damage (i.e. before the traction reaches cohesive strength T_o) follows linearly elastic behavior. Damage initiation refers to the start of stiffness degradation of the cohesive interface [16], and occurs when the stress/traction applied on the interface satisfies certain damage initiation criteria. In this work, a quadratic nominal stress criterion is used for damage initiation, which states that damage starts when a quadratic interaction function involving the nominal stress ratios reaches unity [14]:

$$\left\{ \frac{\langle T_n \rangle}{T_n^o} \right\}^2 + \left\{ \frac{T_s}{T_s^o} \right\}^2 + \left\{ \frac{T_t}{T_t^o} \right\}^2 = 1 \tag{3.3}$$

where, T_n, T_s, and T_t are the tractions on the interface in the normal, the first shear, and the second shear directions, respectively. T_n^o, T_s^o, and T_t^o are the peak values

Fig. 3.2 A bilinear
traction-separation law

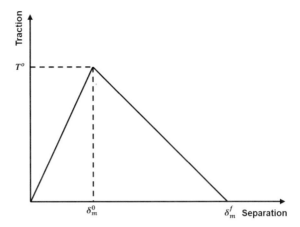

of the nominal stress when the deformation is purely normal to the interface (pure tension), purely in the first shear direction, and purely in the second shear direction, respectively. The symbol $\langle\rangle$ represents the Macaulay bracket, used to signify that a pure compressive stress state does not initiate damage at the interface.

Damage evolution begins once the damage initiation is reached. Damage evolution is characterized by a progressive degradation of the interface stiffness. A scalar damage variable, D, is used to represent the degree of damage. D evolves from 0 to 1 as damage develops [16]. Assuming linear softening during damage evolution, the damage variable D can be expressed as [17, 18]:

$$D = \frac{\delta_m^f \left(\delta_m^{\max} - \delta_m^0\right)}{\delta_m^{\max} \left(\delta_m^f - \delta_m^0\right)} \tag{3.4}$$

$$\delta_m = \sqrt{\left\langle\delta_n^2\right\rangle + \delta_s^2 + \delta_t^2} \tag{3.5}$$

where δ_m is the effective displacement, used to describe the evolution of damage under a combination of normal and shear deformation across the interface; δ_n is the displacement in the normal direction and the Macaulay bracket $\langle\rangle$ represents that a pure compressive displacement does not contribute to the effective displacement; δ_s and δ_t are the shear displacement in the first and second shear directions, respectively; δ_m^f and δ_m^0 are the effective displacement at the complete failure and at the initiation of damage, respectively. $\delta_m^{\max}$ is the maximum effective displacement attained during the loading history.

The interface stress and stiffness of the traction-separation model during damage evolution are functions of the damage variable D [19], as defined in Eqs. (3.6) and (3.7) and schematically illustrated in Fig. 3.3.

$$T_d = (1 - D)\,\bar{T} \tag{3.6}$$

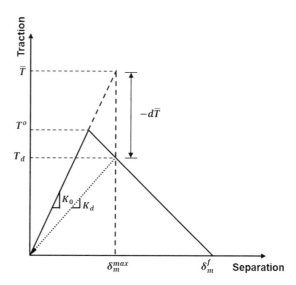

Fig. 3.3 Damage response of the traction-separation law

$$K_d = (1 - D) K_0 \tag{3.7}$$

where T_d is the interface stress at the maximum opening displacement δ_m^{max} attained during the loading history; $\bar{T}$ is the interface stress predicted by the elastic traction-separation behavior for displacement δ_m^{max} without damage; K_d is the stiffness after damage evolution; and K_0 is the initial stiffness before damage.

Fracture propagation is modeled as a result of the damage evolution which can be defined based on fracture energy dissipated during the damage process. The fracture energy is equal to the area under the traction-separation curve (Fig. 3.2) and dependent on the failure modes. In this study, the Benzeggagh-Kenane fracture criterion (Eq. 3.8) is used to define the dependence of fracture energy on failure modes [3, 20]. This criterion states that the mixed-mode failure is interactively governed by the energies required to cause failure in the individual (i.e. normal and two shear) modes.

$$G_n^C + \left(G_s^C - G_n^C\right) \left(\frac{G_S}{G_T}\right)^\beta = G^C \tag{3.8}$$

where $G_S = G_s + G_t$ is the total energy dissipated due to deformations in the first and second shear directions; $G_T = G_n + G_s + G_t$ is the total energy dissipated due to deformations in the normal, the first shear, and the second shear directions; $G^C = G_n^C + G_s^C + G_t^C$ is the total critical fracture energy in the normal, the first shear, and the second shear directions; G_n, G_s and G_t are the energies dissipated due to deformations in the normal, the first shear, and the second shear directions, respectively; G_n^C, G_s^C and G_t^C are the critical energies required to cause failure in the normal, the first shear, and the second shear directions, respectively.

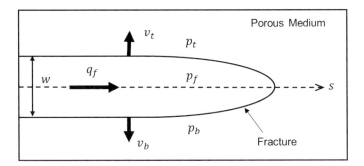

Fig. 3.4 Tangential flow and leak-off of fracture fluid

Upon complete failure of the cohesive interface, i.e. the fracture opens, fluid will flow into the fracture. Both tangential flow along the fracture direction and fluid leak-off from the fracture surface to the surrounding formation may occur, as shown in Fig. 3.4. The continuity equation for fracture fluid flow can be expressed as [21, 22]:

$$\frac{\partial w}{\partial t} + \frac{\partial q_f}{\partial s} + v_t + v_b = 0 \tag{3.9}$$

where w is the fracture width; q_f is the tangential flow rate along the fracture direction; v_t and v_b are the leak-off velocities through the top and bottom surfaces of the fracture.

The tangential flow is modeled as Newtonian fluid flow, and described with the following momentum equation:

$$q_f = -\frac{w^3}{12\mu_f}\frac{\partial p_f}{\partial s} \tag{3.10}$$

where q_f is the tangential flow rate; w is fracture width; μ_f is the fluid viscosity; p_f is the fluid pressure; s the distance along the fracture.

The leak-off (normal flow) is defined as [14]:

$$v_{t,b} = c_{t,b}\left(p_f - P_{t,b}\right) \tag{3.11}$$

where v is the leak-off velocity; c is leak-off coefficient; p_f is the fluid pressure inside the fracture, P is the pore fluid pressure in the porous medium adjacent to the fracture surfaces; the subscripts t and b represent the values corresponding to the top and bottom surfaces of the fractures, respectively.

Rock Deformation and Pore Fluid Flow. The formation rock is assumed to be an isotropic and poroelastic medium. Rock deformation and pore fluid flow are modeled based on the theory of poroelasticity. With the sign convention that tension is positive and compression is negative, the relationship between total stress σ, effective stress σ', and pore pressure p_p, can be expressed as [23]:

$$\sigma = \sigma' - \alpha p_p \boldsymbol{I} \tag{3.12}$$

where $\boldsymbol{I}$ is unit matrix; α is the Biot coefficient.

Stress equilibrium for the solid phase of the porous material is expressed using the principle of virtual work for the volume under its current configuration [22, 24]:

$$\int_V \sigma : \delta\boldsymbol{\varepsilon} \, dV = \int_S \boldsymbol{t} \cdot \delta\boldsymbol{v} \, dS + \int_V \boldsymbol{f} \cdot \delta\boldsymbol{v} \, dV \tag{3.13}$$

where V is the control volume; S is the surface area under surface traction; σ is the total stress matrix; $\delta\boldsymbol{\varepsilon}$ is the virtual strain rate matrix; $\boldsymbol{t}$ is the surface traction vector; $\boldsymbol{f}$ is the body force vector; and $\delta\boldsymbol{v}$ is the virtual velocity vector. This equation is discretized using a Lagrangian formulation for the solid phase, with displacements as the nodal variables. The porous medium is thus modeled by attaching the finite element mesh to the solid phase. Fluid is allowed to flow through these meshes.

A continuity equation is required for fluid flow in the porous medium. The equation equates the rate of change of the total fluid mass in the control volume V to the fluid mass crossing the surface S per unit time, and can be expressed as [16]:

$$\frac{d}{dt}\left(\int_V \rho_f \varphi dV \right) = - \int_S \rho_f \boldsymbol{n} \cdot \boldsymbol{v}_{fp} dS \tag{3.14}$$

where ρ_f is the density of the pore fluid; φ is the porosity of the medium; $\boldsymbol{v}_{fp}$ is the average velocity of the pore fluid relative to the solid phase; $\boldsymbol{n}$ is the outward normal to surface S.

The pore fluid flow in the formation follows Darcy's law as:

$$\boldsymbol{v}_{fp} = -\frac{1}{\varphi g \rho_f} \boldsymbol{k} \cdot \left(\frac{\partial p_p}{\partial X} - \rho_f \boldsymbol{g} \right) \tag{3.15}$$

where $\boldsymbol{g}$ is the gravity acceleration vector; g is the magnitude of gravity acceleration; $\boldsymbol{k}$ is the hydraulic conductivity of the porous medium; p_p is pore pressure; X is a spatial coordinate vector.

3.2.2 Formulation of the Lost Circulation Model

The development of the lost circulation FEM model is described in this section. During drilling, the drilling mud is pumped into the well through the drill pipe and returns to the surface through the annulus (see Fig. 3.1). The formation is assumed to be in a 2D, plane-strain condition. Owing to symmetry, only one half of the formation

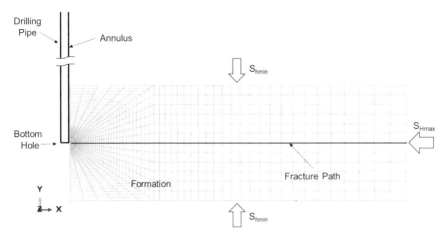

Fig. 3.5 The lost circulation model. The formation is in a plane-strain condition in the horizontal x-y plane. The drilling pipe and annulus are in the z-direction perpendicular to the x-y plane, but they are shown in the y-direction in the figure for better visualization (after [3], with permission from SPE)

is modeled as shown in Fig. 3.5. The maximum and minimum horizontal stresses (S_{Hmax} and S_{hmin}) are in the X- and Y-directions in the horizontal plane, respectively.

The well, consisting of drill pipe and annulus, is modeled as a "U-tube" configuration (Fig. 3.5). For simplicity, the annulus is assumed to have a uniform size (i.e. constant clearance between the drilling pipe and the wellbore wall) along the well depth. In a real situation, the annulus clearance would vary due to multiple casing sizes, which can be readily taken into account in the model once configuration of the casing programs is known. The drill pipe and annulus are discretized using the pipe elements in Abaqus.

A predefined fracture path is assigned in the direction of the maximum horizontal stress from the wellbore wall to the outside of the formation (Fig. 3.5). The fracture path is discretized using a layer of coupled pore pressure cohesive zone elements in Abaqus. To improve the model accuracy, progressively refined elements towards the wellbore are used because significant stress gradients are expected in the wellbore vicinity.

A symmetric boundary condition is defined on the left edge of the model (see Fig. 3.5). The normal displacements of the other three outer boundaries are restricted. Initial field stresses (S_{Hmax} and S_{hmin}) and pore pressure are applied to the whole domain of the model. A constant pore pressure boundary condition (equal to the initial formation pressure) is applied to all the outer boundaries, except the symmetric boundary. The wellbore is filled with drilling fluid. Gravity force is applied to the fluid in the wellbore. The pressure at the wellhead (annulus side) is equal to the atmospheric pressure (assumed to be zero since its small value compared with fluid pressure in the wellbore). A constant pump rate is defined at the top of the drill pipe to simulate the pumping of mud.

Table 3.1 Input parameters for the lost circulation model (after [3], with permission from SPE)

Parameters	Values	Parameters	Values
Formation size	60×20 m	Rock Young's modulus	7 GPa
Formation depth	1000 m	Rock Poisson's ratio	0.2
Wellbore radius	10 cm	Rock permeability	5 mD
Drilling pipe radius	5 cm	Rock porosity	0.25
Annulus clearance	5 cm	Critical tensile strength	0.4 MPa
Initial pore pressure	10 MPa	Fracture toughness	28 J/m^2
Minimum horizontal stress	13 MPa	Leak-off coefficient	5×10^{-9} m/s/Pa
Maximum horizontal stress	15 MPa	Interface stiffness	80 GPa
Pore fluid density	1.0 g/cm^3	Pumping rate	0.36 m^3/min
Drilling mud density	1.3 g/cm^3	Fluid viscosity	1 cp
Gravity constant	10 m/s^2		

During the simulation, it is important to ensure stress equilibrium and fluid continuity between the wellbore, the formation, and the fracture at the bottom of the hole. These requirements are achieved by using the "TIE" constraint and the "PORMECH" constraint available in Abaqus. The "TIE" constraint links the last annulus element node to the formation element nodes on the wellbore wall and the cohesive element node at the fracture mouth. This constraint ensures the pore pressure in the bottom annulus is equal to the pore pressure at the wellbore wall (assuming permeable wellbore wall) and the fracture mouth. The "PORMECH" constraint is used to automatically apply the dynamic fluid pressure at the bottom annulus onto the wellbore wall as a mechanical surface pressure. These two particular constraints, therefore, successfully handle the coupling among the wellbore, the formation, and the fracture.

3.2.3   Results of the Lost Circulation Model

The model described above provides a unique way to model lost circulation during dynamic mud circulation. It can capture the dynamic loss rate, the returned circulation rate, BHP, and lost circulation fracture profiles, which are valuable information for lost circulation prevention, mud optimization, and LCMs selection. Some key modeling results are presented in this section to illustrate the capability of the model. Table 3.1 reports the base-case input parameters used in the modeling.

Using the data reported in Table 3.1, a short period of 100-second mud circulation is simulated because the early-time fluid loss and fracture behavior are of particular

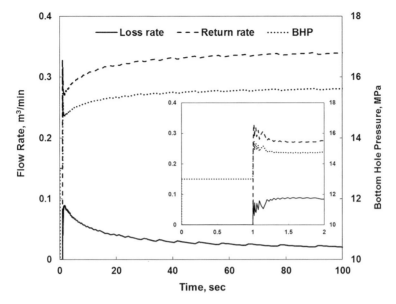

Fig. 3.6 Fluid loss rate, return rate, and BHP during mud circulation. The inset is a zoom-in plot for the first 2-seconds circulation (after [3], with permission from SPE)

interest in the studies of lost circulation and wellbore strengthening. The time developments of mud loss rate, returned circulation rate, and BHP during mud circulation are shown in Fig. 3.6. The results show a BHP of 13 MPa before the start of mud circulation, which is the hydrostatic pressure of the mud column in the annulus. Immediately after the start of circulation, the BHP increases to 15.5 MPa due to the dynamic friction loss in the annulus. This BHP exceeds the fracture initiation pressure, resulting in sudden fracture creation and fluid loss. The BHP decreases to 14.7 MPa with the sudden fracture creation, and then gradually returns to 15.5 MPa with further mud circulation. The fluid loss rate first increases to 0.09 m^3/min (25% pumping rate), and then gradually decreases to 0.02 m^3/min (6% pumping rate). Accordingly, the fluid return rate first decreases to 0.27 m^3/min and then increases to 0.34 m^3/min. The results show that neither the fluid loss nor the BHP are constant during the early-time of lost circulation.

The model can predict real-time fracture geometry, critical information for optimizing particle size distribution (PSD) in wellbore strengthening applications [25]. The time developments of fracture mouth width and fracture length are shown in Fig. 3.7. The results show that the fracture growth is rapid in the early circulation, and then slows down with time. This is more obvious for the growth of fracture mouth width. The fracture mouth width finally reaches a value of about 1.4 mm, and further increase in circulation time will not significantly change the fracture mouth width, while the fracture length will continue growing. However, the fracture will not extend indefinitely. The fracture growth will cease when the BHP is no longer sufficient to

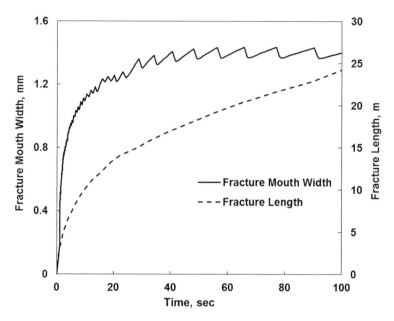

Fig. 3.7 Evaluation of fracture mouth width and fracture length during mud circulation (after [3], with permission from SPE)

overcome the pressure loss along the fracture or an equilibrium between the fluid flow into the fracture and the fluid leak-off from the fracture is established. Oscillating fracture growth behavior is observed, especially for the fracture width. This phenomenon results from intermittent fracture advancement and is also observed in field and laboratory tests [26–30]. The oscillating fracture width may influence the effectiveness of the LCM particles used in wellbore strengthening applications and deserves further investigation. The temporal extent of the fracture (in both width and length) is further illustrated in Fig. 3.8.

It is also interesting to know the stress and pore pressure distributions around the fracture during lost circulation. Figure 3.9 depicts pore pressure distribution at four different mud circulation moments of 1, 5, 10 and 50 s. It can be seen that the pore pressure around the wellbore and near the fracture builds up with circulation time due to fluid diffusion into the formation through the wellbore wall and fracture surfaces. The maximum pore pressure appears in the near-wellbore region. Figure 3.10 shows the distribution of the maximum effective principal stress around the fracture at different moments, with the sign convention of tensile stress as positive and compressive stress as negative. The results show that greater (more tensile) maximum effective principal stress occurs near the wellbore and the fracture during fracture propagation. The greatest value appears in the region close to the wellbore because the relatively larger fluid diffusion in this area leads to larger tensile stress in the formation.

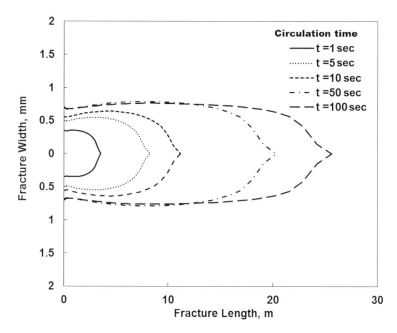

Fig. 3.8 Fracture geometry at different circulation times (after [3], with permission from SPE)

3.3 Modeling Study of Wellbore Ballooning

3.3.1 Wellbore Ballooning Overview

Borehole ballooning is the phenomenon of reversible mud losses and gains during drilling [31]. It has been a major, but not well-understood, problem in the drilling industry. Borehole ballooning is an indicator of a likely subsequent lost circulation event. Failure to control borehole ballooning may result in significant fluid loss and, consequently, increased drilling time and cost [9, 10]. Furthermore, mud gain from a formation in the pump-off period may be diagnosed as a well kick, prompting an increased mud weight to prevent it. This could lead to even worse lost circulation events, especially in high-pressure, high-temperature (HPHT) formations, where the safe drilling margin is narrow and a minor change in mud weight can cause wellbore failure [32]. On the other hand, borehole ballooning also has its valuable aspects. For example, [33] argued that some information collected from borehole ballooning can be used to constrain the fracture gradient of a wellbore.

Borehole ballooning mostly occurs in naturally fractured formations. During mud circulation, high equivalent circulation density (ECD) in the annulus resulting from additional frictional pressure exceeds reopening pressure of the natural fractures, resulting in mud loss into the formation. With pumps off, the annulus pressure falls below the fracture reopening pressure due to the removal of frictional pressure, and

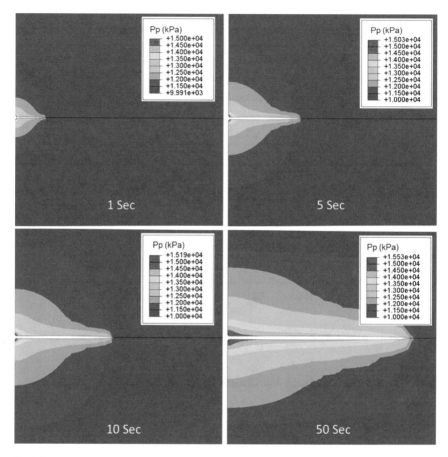

Fig. 3.9 Pore pressure distribution around the fracture during lost circulation (after [3], with permission from SPE)

a sizeable amount of mud flows back into the wellbore. An accurate model capturing this process can aid in understating the mechanisms behind borehole ballooning, distinguishing it from a well kick, and improving mud optimization [8, 34, 35].

For a comprehensive borehole ballooning model, three major physical processes should be taken into account and modeled simultaneously. They are wellbore hydraulics, fracture opening/closing, and deformation of porous formation. There is a very limited number of published borehole ballooning models. To the authors' knowledge, none of them couples these three components.

[11] proposed a mud loss model which assumes mud flows into a non-deformable fracture with a constant aperture and impermeable fracture walls. Similarly, [6] developed a model for mud flow into a non-deformable, infinite radial fracture. These models, neglecting fracture deformation with pressure buildup inside the fracture, may cause underestimation of fluid loss volume [34]. [5] introduced a borehole balloon-

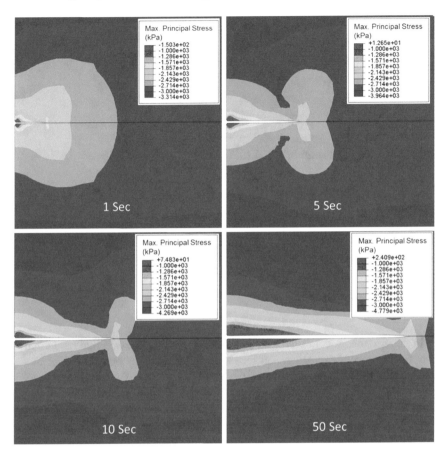

Fig. 3.10 Maximum principal stress distribution around the fracture during lost circulation (after [3], with permission from SPE)

ing model for fluid loss into a fracture of finite length undergoing fracture aperture change with a linear deformation law. Later, they extended the model to fracture deformation with an exponential deformation law [31]. [9] also used a model with a linear fracture deformation law to investigate borehole ballooning with an emphasis on the effect of fracture roughness. Similar models using a linear or exponential deformation law to relate fracture aperture and fluid pressure in the fracture were also presented in [7, 36]. None of these models describes the initiation, propagation, and closure of the fractures based on fracture mechanics theory. Moreover, they do not explicitly model mud circulation in the wellbore as well as rock deformation and pore fluid flow in the bulk formation surrounding the fracture. Therefore, they cannot capture fluid exchanges between the wellbore, fracture, and formation during borehole ballooning events.

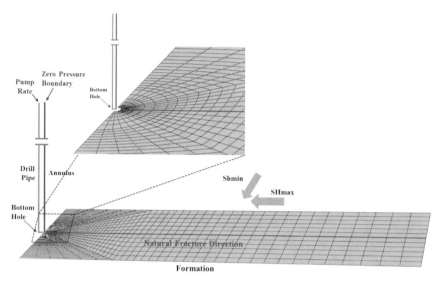

Fig. 3.11 The borehole ballooning model

In this section, a wellbore ballooning model that couples dynamic mud circulation, fracture opening/closing, and formation deformation is presented. The model significantly reduces the shortcomings of previous models. The modeling theory is very similar to that of the lost circulation model in Sect. 3.2. Mud circulation is modeled based on Bernoulli's equation, taking into account both gravity and viscous pressure losses. Fracture behavior is described using the cohesive zone approach. The model captures mud losses into the formation during mud circulation and mud gains from the formation during pump-off period. In addition, the model can provide estimates of time-dependent wellbore pressure, fluid loss/gain rate, and the fracture profile. A numerical example is carried out in this section to illustrate the capabilities of the model.

3.3.2   Formulation of the Wellbore Ballooning Model

Borehole ballooning in a vertical wellbore is considered. The formation is assumed in a 2D, plane-strain condition with unit thickness, as shown in Fig. 3.11. The wellbore is modeled as a 'U-tube' geometry with the left and right side of the U-tube representing drill pipe and wellbore annulus, respectively. During mud circulation, a fluid pump rate is applied to the top of the drill pipe. The pump rate is then reduced to zero in the pump-off period. Owing to symmetry, only one-half of the formation is modeled. The path of a natural fracture is predefined perpendicular to the direction of the minimum horizontal stress and intersecting with the wellbore as shown in Fig. 3.11.

Table 3.2 Input parameters for the borehole ballooning model

Parameters	Values	Parameters	Values
Formation size	60×20 m	Rock Young's modulus	7 GPa
Formation depth	1000 m	Rock Poisson's ratio	0.2
Wellbore radius	10 cm	Rock permeability	5 mD
Drilling pipe radius	5 cm	Rock porosity	0.25
Annulus clearance	5 cm	Cohesive strength	0.4 MPa
Initial pore pressure	10 MPa	Fracture toughness	28 J/m^2
Minimum horizontal stress	13 MPa	Leak-off coefficient	5×10^{-9} m/s/Pa
Maximum horizontal stress	15 MPa	Interface stiffness	80 GPa
Pore fluid density	1000 kg/m^3	Pumping rate	0.36 m^3/min
Drilling mud density	1200 kg/m^3	Mud viscosity	10 cp
Gravity constant	10 m/s^2	Natural fracture length	15 m

A uniform initial pore pressure is applied to the formation. The minimum and maximum horizontal stresses are applied to the directions as shown in Fig. 3.11. A symmetric boundary condition is defined on the left edge of the model. The normal displacements of all the other external boundaries are restricted, and the pore pressure at these boundaries is restricted to the initial pore pressure during the entire simulation. The end nodes of the wellbore annulus are connected to the nodes on the wellbore wall to ensure fluid conservation between the wellbore and the formation. In addition, another special constraint is imposed to assure the external force acting on the wellbore wall is equal to the fluid pressure at the bottom of the annulus during the simulations. The calculations are done using the Abaqus finite-element solver.

3.3.3 Results of the Wellbore Ballooning Model

Contrary to some existing models that only model the fracture itself with a given aperture, the proposed model in this section allows for the hydraulic fracture to grow and close with time-dependent aperture and length controlled by complex interactions between mud circulation, fracture fluid flow, fluid leak-off, and rock deformation. A wealth of information can be provided by the model, including the fluid loss/gain rate, downhole pressure, fracture profile, as well as the pressure and stress distribution in the local area. This section presents some simulation results using input data reported in Table 3.2. For illustration purposes, a mud circulation period of 50 s and a pump-off period of 100 s are considered.

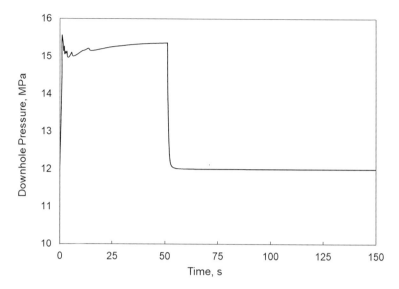

Fig. 3.12 Downhole pressure versus time during the wellbore ballooning event

Figure 3.12 shows the downhole pressure during the borehole ballooning event. After the start of mud circulation, the downhole pressure increases rapidly from a hydrostatic pressure of 12 MPa to a dynamic circulation pressure of 15.5 MPa. This circulation pressure is higher than the reopening pressure of the natural fracture. Therefore, the fracture opens as shown in Fig. 3.13, and consequently, mud flows into the fracture as shown in Fig. 3.14. Note that in Fig. 3.14, the positive rate means mud loss from the borehole into the formation during mud circulation, and the negative rate is the rate of mud gain at the wellhead with pumps off.

Figure 3.12 indicates that the downhole pressure decreases after the initiation of the natural fracture. At the early time of fracture propagation, downhole pressure fluctuations are observed. This downhole pressure behavior is a part of the solution of the simulation, which is quite different from existing models that assume a constant downhole pressure (as an input) during mud loss. The fracture extends to the end of the natural fracture at about 20 s. After that, the downhole pressure, fracture width, and mud loss rate reach relatively constant values with continuing mud circulation. At this period, the mud loss rate is dominated by fluid leak off into formation, rather than by fracture growth.

With pumps stopped, the downhole pressure drops almost immediately to the hydrostatic pressure of 12 MPa. As a result, the fracture aperture decreases to zero, but at a rate slower than the decrease of downhole pressure. With fracture closing, fluid flows out of the wellbore with a gradually decreased rate as shown in Fig. 3.14.

Figure 3.15 illustrates the cumulative fluid volume lost into the formation with time during the circulation and pump-off periods, which is the time integral of the flow rate in Fig. 3.14. The total amount of mud lost into the formation during drilling

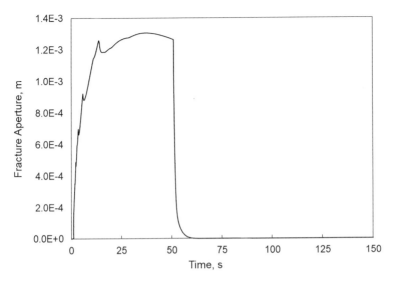

Fig. 3.13 Fracture mouth aperture versus time during the wellbore ballooning event

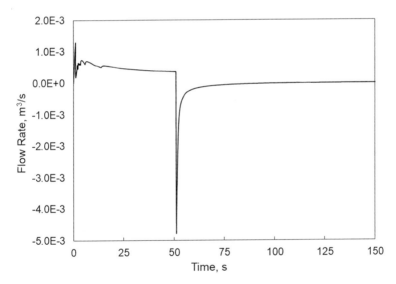

Fig. 3.14 Fluid loss/gain rate versus time during the wellbore ballooning event

period is about 0.025 m³, while the final fluid loss at the end of pump-off period is 0.015 m³, indicating a fluid gain of 0.01 m³ after the stop of drilling. The fluid volume that does not return during pump-off period is the part of fluid filtration into the formation. The results demonstrate that the proposed model can simulate the entire process of a borehole ballooning event.

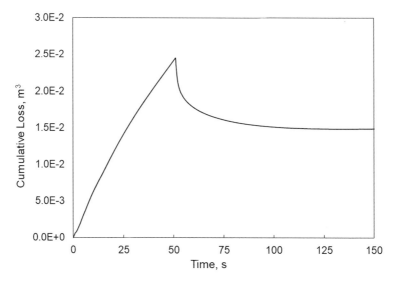

Fig. 3.15 Cumulative fluid loss versus time during the wellbore ballooning event

In summary, modeling borehole ballooning is a challenging endeavor due to strong interactions between wellbore hydraulics, fracture opening/closing, and formation deformation. The numerical model takes into account the non-linear coupling amongst these phenomena. The numerical example indicates that the model is able to capture dynamic fracture growth during mud circulation when downhole ECD is higher than fracture reopening pressure of natural fractures, as well as fracture closure during the pump-off period when downhole pressure is lower. Time-dependent wellbore pressure, fluid loss/gain rate, and fracture profile in borehole ballooning can be obtained. The developed model aids in understanding the mechanisms involved in wellbore ballooning in naturally fractured formations. It can be used to investigate the effects of various operational and in situ parameters on wellbore ballooning, and aids in mud optimization and drilling operations.

References

1. Feng Y, Gray KE (2017) Review of fundamental studies on lost circulation and wellbore strengthening. J Pet Sci Eng 152:511–522
2. Feng Y, Jones JF, Gray KE (2016) A review on fracture-initiation and -propagation pressures for lost circulation and wellbore strengthening. SPE Drilling Completion 31(02):134–144
3. Feng Y, Gray KE (2018) Modeling lost circulation through drilling-induced fractures. SPE J 23(01):205–223
4. Kostov N, Ning J, Gosavi SV, Gupta P, Kulkarni K, Sanz P (2015) Advanced drilling induced fracture modeling for wellbore integrity prediction. In: SPE annual technical conference and exhibition, 28–30 Sept, Houston, Texas, USA. https://doi.org/10.2118/174911-MS

5. Lavrov A, Tronvoll J (2004) Modeling mud loss in fractured formations. In: Abu Dhabi international conference and exhibition, 10–13 Oct, Abu Dhabi, United Arab Emirates. https://doi.org/10.2118/88700-MS

6. Lietard O, Unwin T, Guillot DJ, Hodder MH (1999) Fracture width logging while drilling and drilling mud/loss-circulation-material selection guidelines in naturally fractured reservoirs (includes associated papers 75283, 75284, 81590 and 81591). SPE Drilling Completion 14(03):168–177

7. Majidi R, Miska S, Thompson LG, Yu M, Zhang J (2010) Quantitative analysis of mud losses in naturally fractured reservoirs: the effect of rheology. SPE Drilling Completion 25(04):509–517

8. Mehrabi M, Zeyghami M, Shahri MP (2012) Modeling of fracture ballooning in naturally fractured reservoirs: a sensitivity analysis. In: Nigeria annual international conference and exhibition, 6–8 Aug, Lagos, Nigeria. https://doi.org/10.2118/163034-MS

9. Ozdemirtas M, Babadagli T, Kuru E (2009) Experimental and numerical investigations of borehole ballooning in rough fractures. SPE Drilling Completion 24(02):256–265

10. Ozdemirtas M, Babadagli T, Kuru E (2007) Numerical modelling of borehole ballooning/breathing-effect of fracture roughness. In: Canadian international petroleum conference, 12–14 June, Calgary, Alberta. https://doi.org/10.2118/2007-038

11. Sanfillippo F, Brignoli M, Santarelli FJ, Bezzola C (1997) Characterization of conductive fractures while drilling. In: SPE European formation damage conference, 2–3 June, The Hague, Netherlands. https://doi.org/10.2118/38177-MS

12. Ghalambor A, Salehi S, Shahri MP, Karimi M (2014) Integrated workflow for lost circulation prediction. In: SPE international symposium and exhibition on formation damage control, 26–28 Feb, Lafayette, Louisiana, USA. https://doi.org/10.2118/168123-MS

13. Salehi S (2012) Numerical simulations of fracture propagation and sealing: implications for wellbore strengthening. Doctoral Dissertations, Missouri University of Science and Technology, Rolla, Missouri, USA

14. ABAQUS F (2014) ABAQUS 6.14 documentation. Dassault Syst Provid

15. Churchill SW (1977) Friction-factor equation spans all fluid-flow regimes. Chem Eng 84(24):91–92

16. Zhang GM, Liu H, Zhang J, Wu HA, Wang XX (2010) Three-dimensional finite element simulation and parametric study for horizontal well hydraulic fracture. J Petrol Sci Eng 72(3–4):310–317

17. Camanho PP, Davila CG (2002) Mixed-mode decohesion finite elements for the simulation of delamination in composite materials. NASA/TM-2002-211737. VA, USA: NASA Langley Research Center

18. Turon A, Camanho PP, Costa J, Dávila CG (2006) A damage model for the simulation of delamination in advanced composites under variable-mode loading. Mech Mater 38(11):1072–1089

19. Feng Y, Gray KE (2017) Parameters controlling pressure and fracture behaviors in field injectivity tests: a numerical investigation using coupled flow and geomechanics model. Comput Geotech 87:49–61

20. Zhao P, Santana CL, Feng Y, Gray KE (2017) Mitigating lost circulation: a numerical assessment of wellbore strengthening. J Petrol Sci Eng 157:657–670

21. Li Y et al (2017) Numerical simulation of limited-entry multi-cluster fracturing in horizontal well. J Petrol Sci Eng 152:443–455

22. Yao Y, Gosavi SV, Searles KH, Ellison TK (2010) Cohesive fracture mechanics based analysis to model ductile rock fracture. In: 44th U.S. rock mechanics symposium and 5th U.S. Canada rock mechanics symposium, 27–30 June, Salt Lake City, Utah

23. Biot MA (1941) General theory of three-dimensional consolidation. J Appl Phys 12(2):155–164

24. Wang H (2015) Numerical modeling of non-planar hydraulic fracture propagation in brittle and ductile rocks using XFEM with cohesive zone method. J Petrol Sci Eng 135:127–140

25. Zhong R, Miska S, Yu M (2017) Modeling of near-wellbore fracturing for wellbore strengthening. J Nat Gas Sci Eng 38:475–484

26. Morita N, Black AD, Guh GF (1990) Theory of lost circulation pressure. In: SPE annual technical conference and exhibition, 23–26 Sept, New Orleans, Louisiana. https://doi.org/10.2118/20409-MS

27. Okland D, Gabrielsen GK, Gjerde J, Koen S, Williams EL (2002) The importance of extended leak-off test data for combatting lost circulation. In: SPE/ISRM Rock Mechanics Conference, 20–23 Oct, Irving, Texas. https://doi.org/10.2118/78219-MS

28. Onyia EC (1994) Experimental data analysis of lost-circulation problems during drilling with oil-based mud. SPE Drilling Completion 9(01):25–31

29. Raaen AM, Skomedal E, Kjørholt H, Markestad P, Økland D (2001) Stress determination from hydraulic fracturing tests: the system stiffness approach. Int J Rock Mech Min Sci 38(4):529–541

30. Raaen AM, Brudy M (2001) Pump-in/flowback tests reduce the estimate of horizontal in-situ stress significantly. In: SPE Annual Technical Conference and Exhibition, 30 Sept–3 Oct, New Orleans, Louisiana. https://doi.org/10.2118/71367-MS

31. Lavrov A, Tronvoll J (2005) Mechanics of borehole ballooning in naturally-fractured formations. In: SPE middle east oil and gas show and conference, 12–15 March, Kingdom of Bahrain. https://doi.org/10.2118/93747-MS

32. Helstrup OA, Chen Z, Rahman SS (2004) Time-dependent wellbore instability and ballooning in naturally fractured formations. J Petrol Sci Eng 43(1):113–128

33. Ziegler F, Jones J (2014) Predrill pore-pressure prediction and pore pressure and fluid loss monitoring during drilling: a case study for a deepwater subsalt Gulf of Mexico well and discussion on fracture gradient, fluid losses, and wellbore breathing. Interpretation 2(1):SB45–SB55

34. Bychina M, Thomas GM, Khandelwal R, Samuel R (2017) A robust model to estimate the mud loss into naturally fractured formations. In: SPE annual technical conference and exhibition, 9–11 Oct, San Antonio, Texas, USA. https://doi.org/10.2118/187219-MS

35. Shahri MP, Zeyghami M, Majidi R (2011) Investigation of fracture ballooning and breathing in naturally fractured reservoirs: effect of fracture deformation law. In: Nigeria annual international conference and exhibition, 30 July–3 Aug, Abuja, Nigeria. https://doi.org/10.2118/150817-MS

36. Majidi R, Miska SZ, Yu M, Thompson LG (2008) Fracture ballooning in naturally fractured formations: mechanism and controlling factors. In SPE annual technical conference and exhibition, 21–24 Sept, Denver, Colorado, USA. https://doi.org/10.2118/115526-MS

Chapter 4
Wellbore Strengthening Models

Abstract This chapter presents theoretical models for the evaluation of wellbore strengthening based on plugging/bridging lost circulation fractures using LCMs. An analytical model and a finite-element numerical model are presented. The analytical model is developed based on linear elastic fracture mechanics. It can be used for a fast prediction of fracture pressure enhancement due to fracture bridging, considering near-wellbore stress concentration, in situ stress anisotropy, and different LCM bridge locations. It can also be used to perform quantitative parameter sensitivity analyses to illustrate the undying mechanisms of wellbore strengthening. However, the analytical model cannot provide detailed stress and displacement information local to the wellbore and fracture in wellbore strengthening treatments and it does not consider the porous nature of the formation. To this end, a poroelastic finite-element numerical model is also developed in this chapter. The finite-element model can be used to quantify near-wellbore stress and fracture geometry, before and after bridging fractures. Effects of various parameters are investigated through a comprehensive parametric study using the numerical model. Several useful implications for field applications are obtained based on the parametric study.

4.1 Introduction

Wellbore strengthening operations attempt to "strengthen" the wellbore to prevent or mitigate lost circulation by bridging/sealing lost circulation fractures with LCMs. The ultimate objective is to increase the pressure that a wellbore can sustain without significant fluid loss and widen the drilling mud weight window. While many successful lab experiments and field applications have been reported, the fundamental physics of wellbore strengthening are not thoroughly understood [1, 2]. A lot of disagreement still exists in the drilling industry. There is still a lack of proper mathematic models to quantitatively describe this problem. Thus, for a better understanding of the underlying mechanisms of wellbore strengthening treatments based on bridging lost circulation fractures, both analytical and numerical modeling studies have been introduced in this chapter. The analytical model developed based on the theory

© The Author(s) 2018
Y. Feng and K. E. Gray, *Lost Circulation and Wellbore Strengthening*,
SpringerBriefs in Petroleum Geoscience & Engineering,
https://doi.org/10.1007/978-3-319-89435-5_4

of linear elastic fracture mechanics provides a fast procedure to predict fracture pressure change before and after bridging the fractures, while the numerical model developed based on the finite-element method gives a more detailed description of the distributions of local stress and fracture width in wellbore strengthening.

4.2   A Fracture-Mechanics Model for Wellbore Strengthening

A common practice of wellbore strengthening is utilizing lost circulation materials (LCMs) to bridge the fractures. A number of laboratory and field tests have demonstrated that the fracture breakdown pressure (FBP—wellbore pressure required to advance the lost circulation fractures) can be effectively increased by wellbore strengthening operations [3–6]. Although these tests are very useful for understanding the physics of wellbore strengthening, they are usually expensive and time-consuming to perform and, therefore, only limited information has been obtained from the tests.

Alternatively, theoretical modeling has been used to study wellbore strengthening. Analytical models provide a concise way for wellbore strengthening analysis. A parameter of particular interest in the analytical studies is the FBP of the wellbore, before and after bridging the fractures. Knowledge of FBP is critical for optimizing mud weight and controlling bottom hole pressure during drilling.

Analytical solutions provide equations that directly illustrate the influences of various factors, e.g. in situ stresses, bridging locations, and pore pressure, on FBP. For wellbore strengthening analysis, it is generally important to consider the effect of near-wellbore stress concentration for small fractures within the wellbore vicinity and the effect of pressure drop along the fracture for large fractures extending far beyond the near-wellbore region [7, 8].

Two closed-form analytical models are presented in this chapter to investigate the bridging of small and large fractures, respectively. The models are derived based on linear elastic fracture mechanics and the superposition principle. They can be used to assess FBP before and after bridging the fractures at different locations. A sensitivity study is carried out to evaluate the effects of a number of factors on wellbore strengthening, including the bridging location, field stress, and pore pressure.

4.2.1   Analytical Wellbore Strengthening Model for Bridging Small Fractures

Two short fractures emanating from the wellbore in the direction of maximum horizontal stress (S_{Hmax}) are considered in the small fracture model, as shown in Fig. 4.1. The problem is under a plane strain condition. In wellbore strengthening treatments,

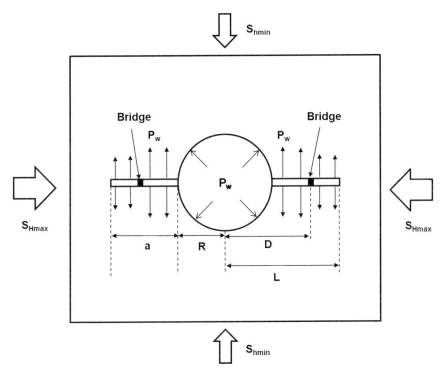

Fig. 4.1 Schematic of the short fracture model (after [9], with permission from Elsevier)

the two fractures are bridged at two symmetric locations. The portion of the fracture from the wellbore wall to the bridge is referred to as invaded zone and the portion from the bridge to the fracture tip is referred to as non-invaded zone. So the larger the invaded zone length, the closer the bridge to the fracture tip.

It is assumed that the bridge can perfectly separate the pressure communication between the invaded and non-invaded zones. Before bridging the fracture, the pressure in the whole fracture is assumed to be equal to the wellbore pressure, considering the negligible pressure drop along the short fracture. After bridging the fracture, the pressure in the invaded zone is equal to the wellbore pressure, while the pressure in non-invaded zone is equal to the formation pore pressure, considering that any overpressure beyond pore pressure in the non-invaded zone will bleed off due to fluid leak-off. Under these assumptions, the stress intensity factor at the fracture tip can be calculated based on the theory of linear elastic fracture mechanics.

The net pressure (equal to fluid pressure in the fracture minus the normal stress acting on the fracture surface) in the invaded zone and non-invaded zone can be calculated as:

$$P_{net\_inv} = P_w - S_{\theta\theta} \tag{4.1}$$

$$P_{net\_noninv} = P_p - S_{\theta\theta} \qquad (4.2)$$

where P_{net_inv} and P_{net_noninv} are the net pressure in the invaded and non-invaded zone, respectively; P_w is wellbore pressure; P_p is formation pore pressure; $S_{\theta\theta}$ is the wellbore tangential stress along the fracture direction, which is equal to the normal stress in the rock acting the fracture surface. $S_{\theta\theta}$ can be approximately determined using the Kirsch solution, assuming the short fracture does not affect the stress concentration around the wellbore [9].

$$S_{\theta\theta} = \frac{1}{2}(S_{Hmax} + S_{hmin})\left(1 + \frac{R^2}{r^2}\right) - \frac{1}{2}(S_{Hmax} - S_{hmin})\left(1 + 3\frac{R^4}{r^4}\right) - P_w\frac{R^2}{r^2}$$

$$(4.3)$$

where S_{Hmax} and S_{hmin} are the maximum and minimum horizontal stresses, respectively; R is wellbore radius; r is the distance from wellbore center to a location along the fracture.

Based on the superposition principle, the fracture-tip stress intensity for the problem shown in Fig. 4.1 can be obtained by calculating the following integrals [9].

$$K_I = \int_R^D \frac{2P_{net\_inv}}{\sqrt{\pi a}}F\left(\frac{r-R}{a}\right)dr + \int_D^L \frac{2P_{net\_noninv}}{\sqrt{\pi a}}F\left(\frac{r-R}{a}\right)dr \qquad (4.4)$$

$$F\left(\frac{r-R}{a}\right) = \frac{1.3 - 0.3\left(\frac{r-R}{a}\right)^{5/4}}{\sqrt{1 - \left(\frac{r-R}{a}\right)^2}}$$

where K_I is the stress intensity factor; D is the distance from wellbore center to the bridge location; a is the fracture length; L is the distance from wellbore center to fracture tip. The first and second integrals on the right-hand side of Eq. (4.4) denote contributions of the net pressures in the invaded and non-invaded zone to the fracture-tip stress intensity factor, respectively.

According to the theory of linear elastic fracture mechanics, fracture extension occurs when the stress intensity factor at the fracture tip reaches the fracture toughness (K_{Ic}) of the material, i.e.

$$K_I = K_{Ic} \qquad (4.5)$$

The fracture breakdown pressure (i.e. the wellbore pressure required to advance the fracture tip) after bridging the fracture can be obtained by combining Eqs. (4.1)–(4.5):

$$P_{fs} = \frac{1}{2} \cdot \frac{1}{F_1 + F_2 - F_4} \cdot K_{IC} + \frac{1}{2} \cdot \frac{F_1 + F_2}{F_1 + F_2 - F_4} \cdot (S_{Hmax} + S_{hmin})$$

$$- \frac{1}{2} \cdot \frac{F_1 + 3F_3}{F_1 + F_2 - F_4} \cdot (S_{Hmax} - S_{hmin}) - \frac{F_4}{F_1 + F_2 - F_4} \cdot P_p \qquad (4.6)$$

.

$$F_1 = \int_R^L G_1(r)\, dr$$

$$F_2 = \int_R^L \frac{R^2}{r^2} G_1(r)\, dr$$

$$F_3 = \int_R^L \frac{R^4}{r^4} G_1(r)\, dr$$

$$F_4 = \int_D^L G_1(r)\, dr$$

$$G_1(r) = \frac{1}{\sqrt{\pi a}} \frac{1.3 - 0.3\left(\frac{r-R}{a}\right)^{5/4}}{\sqrt{1 - \left(\frac{r-R}{a}\right)^2}}$$

where P_{fs} is the breakdown pressure of the bridged fracture in the small fracture model; F_1–F_4 are terms determined by the dimensions of the wellbore-fracture system (including wellbore radius R and fracture length a) and the bridge location D. These terms are referred to as geometry terms in this paper.

Equations (4.6) illustrates that FBP is a function of the stress and pressure applied to the system (S_{Hmax}, S_{hmin} and P_p), the dimensions of the system (R and L), the bridge location (D), and the material property (K_{IC}).

4.2.2   Analytical Wellbore Strengthening Model for Bridging Large Fractures

The large fracture model considers two fractures extending symmetrically from the wellbore wall and perpendicular to the minimum horizontal stress S_{hmin}. The length of the fractures is much larger than the wellbore radius. Under this condition, effects of the presence of the wellbore (i.e. near-wellbore stress concentration) and the far-field stress S_{Hmax} parallel to the fracture can be neglected [10–12]. Thus, the model can be simplified to the problem shown in Fig. 4.2a. Different from the short fracture model, the fluid pressure in the invaded zone in the large fracture model is no longer uniformly equal to the wellbore pressure; instead, it is assumed the pressure drops from the wellbore to the bridge location with a constant gradient. However, the pressure in the non-invaded zone is still assigned equal to formation pore pressure, assuming the bridge perfectly blocks the pressure transmission in the fracture and the overpressure in the non-invaded zone bleeds off into the formation.

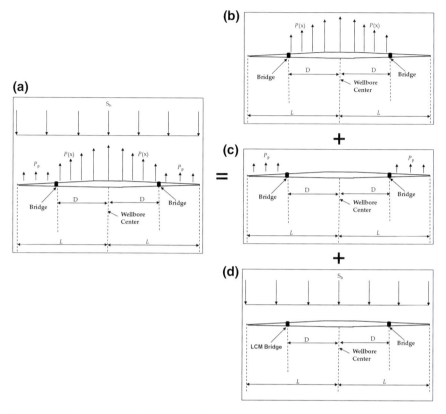

Fig. 4.2 Schematic of the large fracture model (**a**). The model can be decomposed to: the fracture subject to non-uniformly distributed fluid pressure in the invaded zone (**b**), the fracture subject to uniformly distributed fluid pressure equal to the pore pressure in the non-invaded zone (**c**), and the fracture subject solely to the field stress S_h (**d**)

Based on the superposition principle of linear elastic mechanics, the problem shown in Fig. 4.2a can be decomposed into three simpler problems: Problem 1—the fracture subject to a decreasing fluid pressure in the invaded zone from the wellbore to the bridge with a constant gradient (Fig. 4.2b), Problem 2—the fracture subject to uniformly distributed fluid pressure (equal to the pore pressure) in the non-invaded zone (Fig. 4.2c), and Problem 3—the fracture subject solely to the field stress S_{hmin} (Fig. 4.2d).

The pressure distribution in the fracture in Problem 1 with a constant pressure gradient can be defined as:

$$P(x) = P_w - kx, \ (0 \leq x \leq a) \tag{4.7}$$

where x is the distance from wellbore center to a location along the fracture; k is the pressure gradient in the fracture.

Based on linear elastic fracture mechanics, the stress intensity factors at the fracture tip induced by Problems 1, 2 and 3 can be determined as

$$K_{1,1} = \sqrt{\frac{L}{\pi}} \left[2P_w \sin^{-1}\left(\frac{D}{L}\right) - 2k\left(L - \sqrt{(L-D)(L+D)} \right) \right] \qquad (4.8)$$

$$K_{1,2} = \sqrt{\frac{L}{\pi}} \left[2P_p \cos^{-1}\left(\frac{D}{L}\right) \right] \qquad (4.9)$$

$$K_{1,3} = -\sqrt{\pi L}\, S_{hmin} \qquad (4.10)$$

where $K_{1,1}$, $K_{1,2}$ and $K_{1,3}$ are the stress intensity factors of the Problems 1, 2 and 3, respectively.

The stress intensity factor of the original problem shown in Fig. 4.2a can be found by superposing the stress intensity factors given in Eqs. (4.8)–(4.10):

$$K_I = K_{1,1} + K_{1,2} + K_{1,3} \qquad (4.11)$$

By combining Eqs. (4.8)–(4.11) and the fracture criterion Eq. (4.5), the fracture breakdown pressure of the large fracture model can be obtained as:

$$P_{fl} = C_1 K_{IC} + C_2 S_{hmin} + C_3 k + C_4 P_p \qquad (4.12)$$

$$C_1 = \left[2\sqrt{\frac{L}{\pi}} \sin^{-1}\left(\frac{D}{L}\right) \right]^{-1}$$

$$C_2 = \pi \left[2\sin^{-1}\left(\frac{D}{L}\right) \right]^{-1}$$

$$C_3 = \left(L - \sqrt{L^2 - a^2} \right) \left[\sin^{-1}\left(\frac{D}{L}\right) \right]^{-1}$$

$$C_4 = \cos^{-1}\left(\frac{D}{L}\right) \left[\sin^{-1}\left(\frac{D}{L}\right) \right]^{-1}$$

where P_{fl} is the fracture breakdown pressure of the large fracture model; C_1–C_4 are geometry terms determined by the length of the fracture (L) and the bridge location (D).

Equation (4.12) illustrates that fracture breakdown pressure of the large fracture model are functions of the minimum horizontal stress (S_{hmin}), pore pressure (P_p), fracture length (L), bridge location (D), pressure gradient in the fracture (k) and the material property (K_{IC}). Note that the terms of wellbore radius R and maximum horizontal stress S_{Hmax} are not involved in the equation because the large fracture model ignores the effects of the wellbore and the maximum horizontal stress parallel to the fracture.

The small-fracture wellbore strengthening model presented in Sect. 4.2.1 neglects the pressure drop along the fracture; while the large-fracture model described in this section neglects the influence of near-wellbore stress concentration. It is suggested that the small fracture model should be used when the fracture length is less than two wellbore radii, while the large fracture model should be used for fractures longer than two wellbore radii.

4.2.3 Results of the Small Fracture Model

Results of the small-fracture wellbore strengthening models are presented in this section. Effects of far-field stress (anisotropy), bridge location, and pore pressure are investigated. The results are analyzed in term of fracture breakdown pressure.

The base input data for the small fracture model are summarized in Table 4.1. The industry's 6-in fracture length assumption is adopted for the analysis [8, 13, 14]. In each of the following sensitivity analysis, the parameters are kept identical to those in Table 4.1 unless otherwise stated. Fracture breakdown pressure calculated from Eq. (4.6) for various load conditions and bridge locations is reported below.

Figure 4.3 shows the fracture breakdown pressure with various far-field stress anisotropies and bridges locations. The bridge location is denoted by the ratio of the invaded zone length to the fracture length, which increases from 0 to 1 as the bridge moves from the fracture mouth to the fracture tip. The case with bridge location at the fracture tip (i.e. a ratio of 1) is identical to the situation without bridging the fracture. The stress anisotropy is denoted by the ratio of S_{Hmax}/S_{hmin}, which is increased from 1 to 2 by changing the value of S_{Hmax} while keeping S_{hmin} constant at 3000 psi. Figure 4.3 shows the breakdown pressure increases quickly as the invaded zone length decreases (i.e. the bridge location moves closer to the

Table 4.1 Base input parameters used in the proposed model

Parameter	Unit	Value
Wellbore radius (R)	inch	6
Fracture length (a)	inch	6
Minimum horizontal stress (S_{hmin})	psi	3000
Maximum horizontal stress (S_{Hmax})	psi	3600
Wellbore pressure (P_w)	psi	4000
Pore pressure (P_p)	psi	1800
Fracture toughness (K_{IC})	psi-in$^{0.5}$	2000

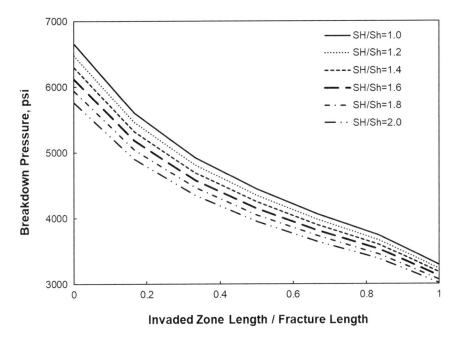

Fig. 4.3 Fracture breakdown pressure for different stress anisotropies with various bridge locations (after [9], with permission from Elsevier)

fracture mouth at wellbore all). This result implies that the best location to bridge the fracture in wellbore strengthening is the region close to fracture mouth if the bridge can effectively isolate the pressure commination ahead and behind it. Figure 4.3 also illustrates that the breakdown pressure decreases with the increase of far-field stress anisotropy. This is because a larger field stress anisotropy results in a less compressive stress acting normally on the fracture in the near-wellbore region, thus a smaller wellbore pressure is required to extend the fracture. However, compared with bridge location, the effect of field stress anisotropy on breakdown pressure is relatively smaller, and the breakdown pressure becomes less sensitive to stress anisotropy with the increase of the invaded zone length.

Figure 4.4 shows the breakdown pressure for different pore pressures with various bridge locations. The variation of pore pressure is denoted by the ratio of pore pressure P_p to the constant S_{hmin} of 3000 psi. The results show that, before bridging the fracture (i.e. the ratio of invaded zone length to fracture length is equal to 1), the breakdown pressure is independent of pore pressure and equal to the value of S_{hmin}. However, the breakdown pressure becomes very sensitive to pore pressure after bridging the fracture. With the increase of P_p/S_{hmin}, the breakdown pressure decreases. This result provides an important implication for the field applications of wellbore strengthening. As aforementioned, two well-known scenarios of lost circulation are drilling through depleted reservoirs and overpressured deepwater formations. The former scenario

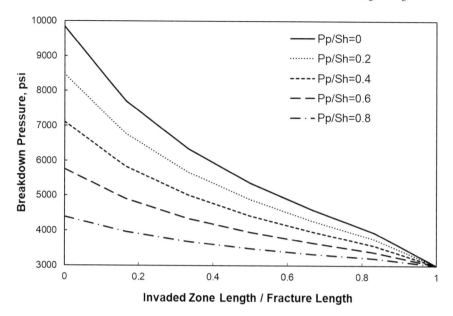

Fig. 4.4 Fracture breakdown pressure for different pore pressures with various bridge locations (after [9], with permission from Elsevier)

usually has a relatively lower P_p/S_{hmin} value due to pore pressure drop, while the latter scenario usually exhibits a larger P_p/S_{hmin} value because of the abnormally high formation pressure and the relatively low field stress caused by the column of seawater. Therefore, according to the result in Fig. 4.4, the wellbore strengthening treatments based on bridging the lost circulation fractures should be more effective in depleted reservoirs as compared to overpressued deepwater formations.

4.2.4 Results of the Large Fracture Model

In the above, results have been given for the small fracture model with a fracture length of 6 in. In this section, sensitivity study results using the large fracture model for a lost circulation fracture of 100 in. are presented. Different from the small fracture model, a pressure gradient in the invaded zone is taken into account in the large fracture model, considering the relatively large viscous pressure dissipation along the fracture. However, the pressure in the non-invaded zone remains equal to the pore pressure, assuming that the drilling fluid does not reach this zone after bridging and the overpressure dissipates into the formation with fluid leak-off. It is assumed that the fracture can be bridged in the region between its midpoint and tip because it is generally very difficult to bridge a large fracture near the fracture

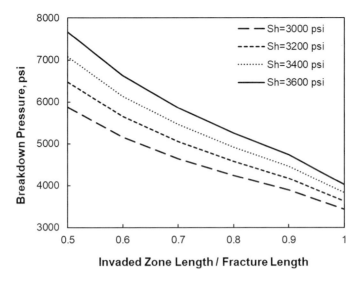

Fig. 4.5 Fracture breakdown pressure for different far-field stress with various bridge locations

mouth [10, 15]. The input parameters used in this section are the same as those reported in Table 4.1, except the fracture length. Since the trends for the effects of the corresponding parameters in the large fracture model are very similar to those in the small fracture model, the results in this section are described very briefly.

Figure 4.5 shows the effect of S_{hmin} on fracture breakdown pressure with different bridge locations. When there is no bridge, the breakdown pressure is slightly higher than S_{hmin}. Similarly to the results of the small fracture model, the existence of a bridge in the fracture contributes to maximize breakdown pressure. Because S_{hmin} is the normal stress acting to close the fracture, the results show that the larger the S_{hmin}, the higher breakdown pressure is required to propagate the fracture, as expected.

Figure 4.6 shows the effect of pore pressure in the large fracture model. The results again show that if P_p is much smaller than S_{hmin}, the breakdown pressure can be significantly enhanced after bridging the fracture. However, if P_p approaches S_{hmin}, the increase in breakdown pressure is much more modest.

Figure 4.7 shows the effect of pressure gradient in the fracture. It can be seen that the larger the pressure gradient, the higher breakdown pressure in the wellbore that is required to extend the fracture. The case without considering pressure gradient predicts a lower limit of breakdown pressure. The breakdown pressure is more sensitive to pressure gradient when the invaded zone length is large due to the associated larger amount of pressure drop.

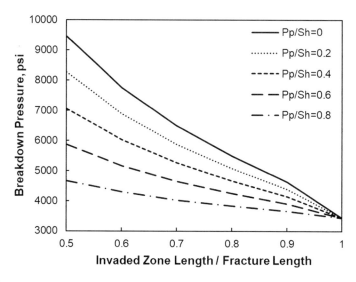

Fig. 4.6 Fracture breakdown pressure for different pore pressure with various bridge locations

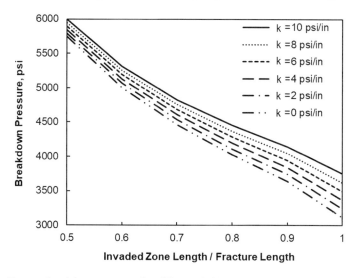

Fig. 4.7 Fracture breakdown pressure for different fluid pressure gradient in the fracture with various bridge locations

4.2.5 Discussions

The results of the small and large fracture models demonstrate that bridging the fracture can effectively maximize the breakdown pressure. However, this conclusion is only valid when the problem satisfies the following two prescribed conditions:

- the LCM bridge is a perfect plug with zero permeability which separates the fluid flow across it completely;
- the formation is relatively permeable and thus any overpressure beyond the pore pressure in the non-invaded zone can dissipate due to fluid leak-off into the formation.

A permeable or leaking LCM bridge will compromise the result of wellbore strengthening [7, 16]. In order to form a qualify LCM bridge in the fracture, filtrate loss from fracture surfaces to the formation is usually required [13, 17–19]. The second prescribed condition implies that the proposed models are more applicable to sandstone, and to a lesser extent, shale formations. The effectiveness of wellbore strengthening treatments can be strongly limited by the low permeability of shale for two reasons:

- the overpressure in the non-invaded zone is difficult to dissipate
- the slow filtration loss restricts the development of LCM bridge [21].

To solve lost circulation problems in low-permeability shales, chemical strategies have been used to strengthen the wellbore, either by changing chemical composition of the formation [20] or by forming chemical sealants in the fracture [21].

It is worth noting the limitations of the analytical models. The models are development based on linear elasticity assumption, without considering the poroelasticity effect. The fractures have fixed length and fluid flow within the fracture is not explicitly modeled. The LCM bridge has zero permeability and the location is prescribed; the transportation, aggregation, and bridging of LCMs within the fracture are not explicitly modeled. Numerical models may have the capabilities to take into account one or more of these additional features. In the following section, a numerical study on wellbore strengthening is presented.

4.3   A Numerical Model for Wellbore Strengthening

While the analytical model described in the above section provides a fast procedure to predict fracture pressure change before and after fracture bridging, it cannot provide detailed stress and displacement information local to the wellbore and fracture in wellbore strengthening treatments. Moreover, it does not consider the poromechanical effect of the formation rock. However, numerical approaches, such as the finite-element method, can be applied to obtain detailed information about the evolution of local stress and deformation in wellbore strengthening operations.

To better simulate wellbore strengthening treatment, a poroelastic finite-element numerical model, considering fluid flow and fluid leak-off is developed. The model quantifies near-wellbore stress and fracture geometry, before and after bridging fractures. Effects of various parameters on wellbore strengthening can be investigated using the model. Although the numerical model cannot directly provide a solution for maximum sustainable pressure of the wellbore in wellbore strengthening analysis, there is no doubt that, from the points of view of both continuum mechanics and fracture mechanics, increasing hoop stress—and, hence, stress acting on closing the fracture—will facilitate the prevention of fracture growth, thus achieving wellbore strengthening. The link between enhancing hoop stress and increasing maximum sustainable pressure of the wellbore has also been discussed in detail in a series of papers [13, 17, 18, 21–23].

4.3.1 Development of the Numerical Wellbore Strengthening Model

Wellbore strengthening treatments for a vertical wellbore are considered. The wellbore is assumed to be in a plane-strain condition. Owing to symmetry, only one quarter of the wellbore is used in the finite-element numerical analysis, as shown in Fig. 4.8. Wellbore radius is 4.25 in. The length and width of the quarter model are 40.25 in.

A pre-existing fracture is assumed on the wellbore as shown in Fig. 4.8, with its face aligned with the maximum horizontal stress (S_{Hmax}) or X-axis direction. The fracture length of 6 in. is used so as to be consistent with previous work by other investigators [13, 14, 24, 25]. The fracture is plugged with LCMs as shown in Fig. 4.8. Through this section, the plug formed inside the fracture by LCMs across the fracture width is called an LCM bridge. The LCM bridge is assumed to be a "perfect" bridge, i.e., a rigid body with zero permeability. So there is no fluid flow across the bridge. To create the effect of bridging the fracture, the velocity in the Y direction at the bridging location is set equal to zero. The quadrant angle is 0 in the X-axis, and increases to 90° around the quarter wellbore.

Symmetric boundary conditions are applied to the left and top boundaries of the model. The maximum horizontal stress (S_{Hmax}) along the X-direction and minimum horizontal stress (S_{hmin}) along the Y-direction are applied to the right and bottom outside boundaries, respectively. Wellbore pressure is applied to the inner wall of the wellbore. Pressure in the fracture is equal to wellbore pressure before bridging the fracture with LCM. After bridging the fracture, pressure in the fracture ahead of the LCM bridge (from wellbore to the bridge) is still equal to wellbore pressure. However, pressure behind the LCM bridge (from the bridge to fracture tip) is set equal to pore pressure, because the pressure in this region will drop to formation pore pressure with fluid leaking off into the formation and no continuous fluid supply from wellbore due to the impermeable bridge. Fluid leak-off velocities are applied on the wellbore wall and fracture faces to simulate fluid leak-off.

Table 4.2 provides the input parameters for the finite-element numerical simulations in the following sections.

- Total size of the model is about ten times the wellbore size in order to eliminate boundary effects on near-wellbore stress and strain states.
- Formation rock properties are selected for a typical sandstone.

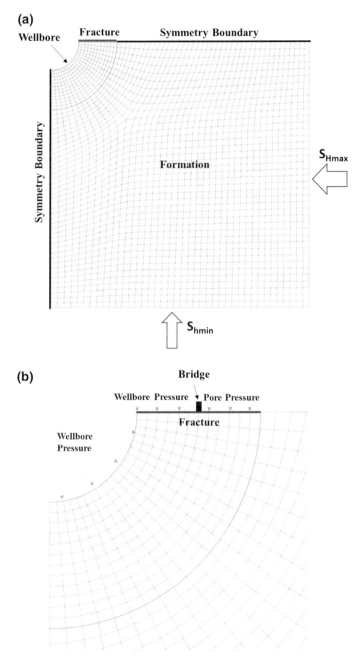

Fig. 4.8 The wellbore strengthening model. **a** Geometry and boundary conditions of the model; **b** detailed fracture process zone (after [9], with permission from Elsevier)

Table 4.2 Input parameters for the finite-element model (after [2], with permission from Elsevier)

Parameter	Values	Units
Model length	40.25	inches
Model width	40.25	inches
Wellbore radius (R)	4.25	inches
Young's modulus (E)	2×10^6	psi
Poisson's ratio (υ)	0.2	
Minimum horizontal stress (S_{hmin})	3000	psi
Maximum horizontal stress (S_{Hmax})	1–1.5 S_{min}	psi
Wellbore pressure (P_w)	4000	psi
Pressure in fracture before bridging (P_{fo})	4000	psi
Pressure ahead of bridge after bridging (P_{fa})	4000	psi
Pressure behind of bridge after bridging (P_{fb})	1800–4000	psi
Fracture length (a)	6	inches
Initial pore pressure (P_p)	1800	psi
Permeability	0.0023	in/min
Void ratio	0.3	
LCM bridge location away from wellbore	0.5, 2.0, 3.5, 5.0	inches

- Maximum and minimum horizontal stresses with different stress contrasts, from 1 ($S_{Hmax}/S_{hmin} = 1$) to 1.5 ($S_{Hmax}/S_{hmin} = 1.5$), are used to investigate the effect of stress anisotropy.
- Different pressures behind the bridge P_{fb} are used to simulate the sealing capacity of the LCM bridge, from complete sealing ($P_{fb} = P_p = 1800$ psi) to no sealing ($P_{fb} = P_p = 40000$ psi).
- Various LCM bridge locations are also selected for parametric sensitivity studies.

In the following, results from finite-element numerical simulations using the input parameters in Table 4.2 are presented. Hoop stress around the wellbore and along the fracture, and fracture width are investigated utilizing the list of influential parameters.

4.3.2 Hoop Stress and Fracture Width

Using the finite-element model described above, hoop stress states in the vicinity of the wellbore and the fracture are analyzed for various combinations of influential parameters. Figure 4.9 shows the hoop stress state before and after bridging the fracture with LCM. The bridge location is 2 in. away from the fracture mouth. Throughout this section, negative and positive stress values mean compressive and tensile stresses, respectively. It is clear that, before bridging, the fracture tip is under tension and near-wellbore rock is under compression. However, after bridging the

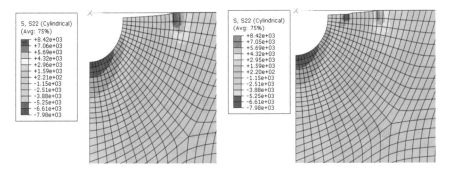

Fig. 4.9 Hoop stress distribution before (left) and after (right) bridging the fracture in remedial wellbore strengthening treatment (after [2], with permission from Elsevier)

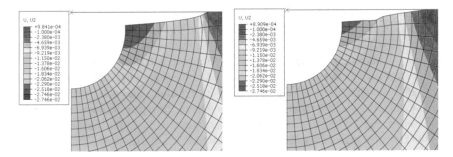

Fig. 4.10 Vertical displacement distribution in the model before and after bridging the fracture in wellbore strengthening (after [2], with permission from Elsevier)

fracture, there is a compressive stress increase area near the bridging location, meaning the fracture is more difficult to open; whereas the tensile stress near the fracture tip decreases, meaning the fracture is more difficult to propagate. After bridging the fracture in wellbore strengthening, lost circulation is less likely to continue.

Figure 4.10 shows vertical displacement perpendicular to the fracture face, before and after bridging the fracture using LCM. The bridge location is 2 in. away from the wellbore wall. The displacement magnitude along the fracture face is the half width of the fracture. The minus sign in Fig. 4.10 means the fracture opening displacement is opposite to the direction of the Y-axis. The blue and red colors indicate larger and smaller fracture opening displacement (or fracture width), respectively. Fracture width decreases after bridging the fracture, especially in the region behind the bridge location.

Note that in the fracture width plots in the following sections, a minus sign is still used. But this sign only means the direction of fracture opening displacement, not a negative fracture width.

4.3.3   Effect of Horizontal Stress Contrast S_{Hmax}/S_{hmin}

Hoop stress on the wellbore wall from $0°$ to $90°$, with different horizontal stress contrasts, for scenarios without fracture, with unbridged fracture, and with bridged fracture, are computed and shown in Figs. 4.11a–c, respectively. For wellbore without fracture, hoop stress on the wellbore within $30°$ to X-axis (S_{Hmax} direction) is tension. With increase in horizontal stress contrast, the tensile stress increases. When there is a fracture created as shown in Fig. 4.11b, the tensile stress on the wellbore wall close to fracture changes to compression. Figure 4.11c is the hoop stress after bridging the fracture at 2 in. away from the wellbore. It's not easy to see the stress differences between wellbore with unbridged fracture and wellbore with bridged fracture from Figs. 4.11b and c. To facilitate the observation, hoop stresses in a single case with $S_{Hmax}/S_{hmin} = 1.3$ before and after bridging the fracture are compared in Fig. 4.11d. It is clearly shown that compressive hoop stress on the wellbore wall increases in the area near the fracture mouth from 0 to $45°$ and decreases in the area beyond $45°$. This is because after bridging the fracture, fluid pressure behind the bridging point will decrease, as a result the fracture will try to close. The healing of the fracture will stretch the formation around it, leading to an increased tension (decreased compression). However, a rigid bridge restrains the healing of the fracture portion near the fracture mouth, resulting in a locally increased compression; in the far area beyond $45°$ increased tension still occurs due to the overall healing behavior of the fracture. Increased compression near the fracture means that the fracture is less likely to be opened after bridging. Decreased compression beyond $45°$ means bridging the fracture actually weakens this wellbore portion and new fractures may generate here with increased wellbore pressure.

Figures 4.12a, b and c show hoop stress along the facture faces with different horizontal stress contrast for scenarios without fracture, with unbridged fracture, and with bridged fracture, respectively. The horizontal axis is the distance away from the wellbore wall along the fracture direction. Before fracture creation, as shown in Fig. 4.12a, hoop stress along the fracture direction is tensile in the near wellbore region, and becomes compressive with increase of distance away from the wellbore wall. The higher the horizontal stress contrast, the larger the tensile stress and tensile area. However, the near-wellbore tensile hoop stress becomes compressive when a fracture is created, as illustrated in Fig. 4.12b; whereas the near-fracture-tip compressive stress becomes tensile. Figure 4.12b also shows horizontal stress contrast has negligible influence on hoop stress along fracture faces. Note that in this study, different S_{Hmax} values are utilized to change horizontal stress contrast, whereas S_{hmin} is kept as a constant value. Figure 4.12c is the stress after bridging the fracture. Horizontal stress contrast still has negligible influence on hoop stress along the fracture. But there is a significant compression increase near the bridge location at 2 in. away from wellbore wall. Hoop stresses along fracture, with horizontal stress contrast equal to 1.3, before and after wellbore strengthening are compared in

Fig. 4.11 Hoop stress on wellbore wall for different horizontal stress contrasts: **a** without fracture, **b** with unbridged fracture, **c** with bridged fracture, and **d** comparison of hoop stresses before and after bridging for horizontal stress contrast equal to 1.3 (after [2], with permission from Elsevier)

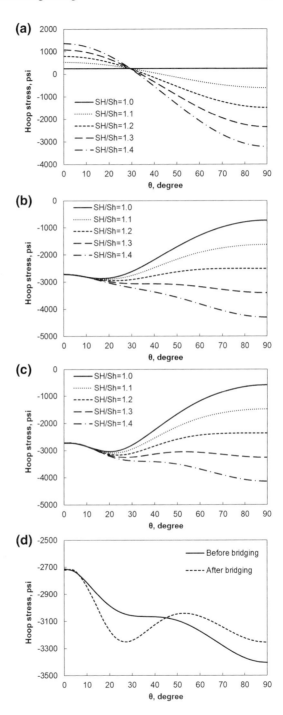

Fig. 4.12 Hoop stress along fracture face for different horizontal stress contrasts: **a** without fracture, **b** with unbridged fracture, **c** with bridged fracture, and **d** comparison of hoop stresses before and after bridging the fracture for horizontal stress contrast equal to 1.3 (after [2], with permission from Elsevier)

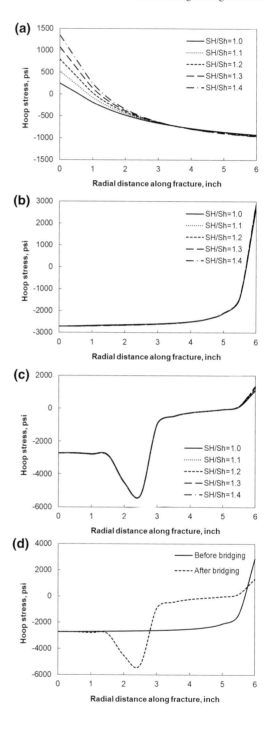

Fig. 4.13 Fracture half-width distribution for different horizontal stress contrasts: **a** before bridging the fracture, **b** after bridging the fracture, and **c** comparison of fracture half-widths before and after bridging the fracture for horizontal stress contrast equal to 1.3 (after [2], with permission from Elsevier)

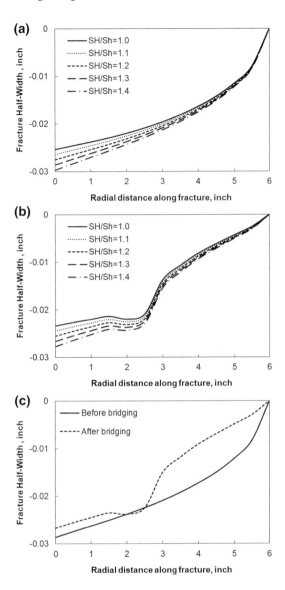

Fig. 4.12d. It is clearly indicated that bridging the fracture will significantly increase the compressive hoop stress near the LCM bridge location, which makes the fracture harder to reopen.

Figures 4.13a and b show fracture half-width distribution along the fracture length with different horizontal stress contrasts with unbridged and bridged fracture, respectively. The bridge location is 2 in. away from the wellbore. For both cases, with increase in horizontal stress contrast, fracture half-width increases, especially in the area close to the wellbore. Figure 4.13b shows that horizontal stress contrast has a very small effect on fracture width behind the LCM bridge after strengthening. Fracture widths before and after bridging the fracture for a particular case with horizontal stress contrast equal to 1.3 are compared in Fig. 4.13c. The results shows that the fracture width behind the LCM bridge has a significant decrease after bridging the fracture, which means the fracture is trying to close after the strengthening operation. However, fracture width experiences a much smaller decrease ahead of bridge location, likely due to the relatively higher fluid pressure in this part of the fracture.

4.3.4 Effect of LCM Bridge Location

Figure 4.14 shows hoop stress on the wellbore wall before and after bridging the fracture at 0.5, 2 and 5 in. away from wellbore. When the LCM bridge is close to the wellbore wall, e.g. the 0.5-in. case, there is a dramatic compressive hoop stress increase on the wellbore wall near the fracture. However, with the bridging location further away from the fracture, e.g. the 2.0-in. case, there is less hoop stress increase. For a bridge at 5.0 in., there is no clear hoop stress change—the stress curve in Fig. 4.14 overlaps with the one before bridging the fracture. The result illustrates that, from a hoop stress enhancement point of view, the best place to bridge the fracture is at the fracture mouth.

Figure 4.15 shows hoop stress along the fracture face with different bridging locations. When the bridge is very close to the wellbore wall, e.g. the 0.5-in. case, there is a dramatic compressive hoop stress increase around the bridge location. However, with the bridge is away from the wellbore wall, the increase of compressive

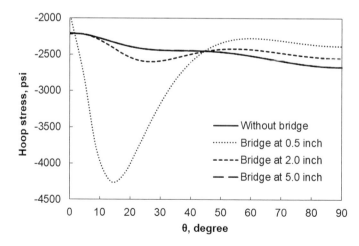

Fig. 4.14 Hoop stress on wellbore wall for different bridging locations (after [2], with permission from Elsevier)

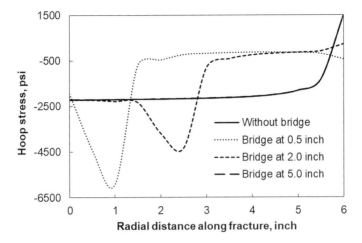

Fig. 4.15 Hoop stress along fracture face for different bridge locations (after [2], with permission from Elsevier)

hoop stress becomes smaller. There is almost no hoop stress change while bridging the fracture near the fracture tip. For example, for the case with bridge at 5.0 in. away from wellbore wall, the stress after bridging is almost the same as that before bridging, indicated as two overlapping curves in Fig. 4.15. These results also demonstrate that the best place to bridge the fracture is near the wellbore wall, and it is important to determine pre-existing fracture width for designing LCM particle size distribution and optimizing bridging location.

Fracture half-widths before bridging the fracture and after bridging it at 0.5, 2.0, 3.5 and 5.0 in. away from the wellbore wall are computed and shown in Fig. 4.16. When the fracture is bridged near the wellbore wall, for example the 0.5-in. case, the fracture experiences a larger decrease in its width, compared with bridging away from the wellbore. For a bridge at 5.0 in., there is no clear width change—the width curve in Fig. 4.16 overlaps with that before bridging the fracture. Since the objective of wellbore strengthening is to prevent fracture opening and propagation, Fig. 4.16 also shows that the best place to bridge the fracture is at the fracture mouth.

4.3.5   *Effect of Pressure Behind LCM Bridge (Bridge Permeability)*

For the numerical studies above, the bridge is assumed to be impermeable and therefore prevents pressure communication across the bridge. It is assumed that pressure behind the LCM bridge will decline to formation pressure as the fluid leaks off. However, in real situations, the bridge is likely not completely impermeable in real situations, and fluid can flow across the bridge due to pressure differential. Depend-

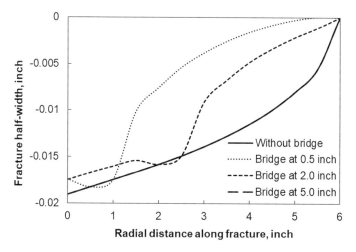

Fig. 4.16 Fracture half-width distribution for different LCM bridge locations (after [2], with permission from Elsevier)

ing on fluid penetration across the bridge or the permeability of the bridge, pressure behind the LCM bridge can vary from formation pressure (perfectly impermeable bridge) to wellbore pressure (fully permeable bridge). Figure 4.17 shows hoop stress along the fracture faces for pressure behind the LCM bridge varying from formation pressure, 1800 psi, to wellbore pressure, 4000 psi. The higher the pressure behind the LCM bridge, the smaller the compression near the bridge location and the higher the tension at the fracture tip. This lessens the effectiveness of wellbore strengthening. The results in Fig. 4.17 illustrate the importance of forming a low permeability bridge.

Figure 4.18 shows the fracture half-width distribution for pressure behind the LCM bridge varying from formation pressure, 1800 psi, to wellbore pressure, 4000 psi. The lower the pressure behind the LCM bridge, the smaller the fracture width behind the LCM bridge. This, again, means the less permeable the LCM bridge, the more effective the strengthening operation.

4.3.6 Field Implications of the Numerical Study

For effectively strengthening a wellbore, it is necessary to understand the fundamentals of lost circulation and wellbore strengthening. Because the fracture growth/closure behavior and corresponding stress and dimension changes are difficult to measure in the field, numerical simulation is especially useful for studying these processes and revealing useful information for field applications. Regarding hoop stress as an important consideration in wellbore strengthening, the following field implications are indicated from the above finite element modeling analysis.

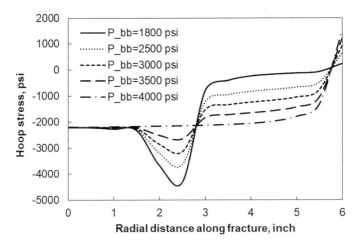

Fig. 4.17 Hoop stress along fracture face for different pressures behind LCM bridge, varying from formation pressure of 1800 psi to wellbore pressure of 4000 psi (after [2], with permission from Elsevier)

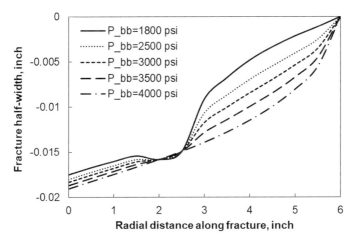

Fig. 4.18 Fracture half-width distribution for different pressure behind LCM bridge, varying from formation pressure 1800 psi to wellbore pressure 4000 psi (after [2], with permission from Elsevier)

- Hoop stress around the wellbore can be substantially increased by bridging the fracture near the wellbore wall. Bridging the fracture near the fracture tip does little to increase hoop stress and strengthen the wellbore. It is important to correctly predict fracture width and optimize LCM size distribution in wellbore strengthening design.
- The LCM plug should have no or very low permeability in order to effectively block wellbore fluid penetration and so that fluid in the fracture tip region can freely dissipate into the nearby permeable formation. This ultimately leads to

fluid pressure in the fracture tip region becoming equal to formation pore pressure, resulting in a closed fracture. The permeability of the bridging plug is a property as important as the bridge strength.

- After effectively bridging the fracture, the hoop stress increases locally on the wellbore wall near the fracture mouth. If the wellbore pressure increases further, subsequent fractures may initiate on the wellbore in some places away from the fracture mouth rather than in the near fracture region.

- With bridging the fracture near its mouth, a substantial hoop stress increase can be achieved. But the hoop stress increase is limited only to the region near the bridging location; after this location, hoop stress decreases quickly to tension. Therefore, it is important to evaluate the potential of hoop stress increase for wellbore strengthening to make sure the downhole pressure or ECD is less than the maximum hoop stress that can be obtained. Wellbore strengthening operations do not work when bridging the fracture far away from its mouth from the point of view of hoop stress enhancement.

- It is well-known from fracture mechanics that the longer the fracture the less net pressure (difference between fluid pressure inside the fracture and field stress acting on fracture faces) is needed to propagate the fracture. However, in a real field situation, the fracture length is hard to predict. However, from these simulation results the best place to bridge the fracture, from a point of view of hoop stress enhancement, is at its mouth or entrance on the wellbore wall. This simplifies the problem, i.e., bridge the fracture mouth, then fracture length is of much less consequence.

4.4 Some Experimental Results of Wellbore Strengthening

In laboratory and field experimental studies of wellbore strengthening, repeated leak-off tests are commonly used to investigate the effectiveness of wellbore strengthening treatments. By performing repeated leak-off tests in a wellbore with and without LCMs in the injection fluid, the fracture pressure for strengthened and un-strengthened wellbore can be compared. In this section, a few selected experimental results of wellbore strengthening are presented.

Figure 4.19 compares the fracture pressure of a wellbore without and with wellbore strengthening treatment in one experimental study of the DEA 13 project [6]. The experiment was conducted on $30 \times 30 \times 30$ in sandstone blocks with a 1.5 in. vertical borehole. The maximum and minimum horizontal stresses applied to the block are 2500 and 1800 psi, respectively. A leak-off test with three cycles is conducted. In the first cycle, oil-based mud with 40-lb/bbl LCMs (calcium carbonates) was injected to fracture the borehole until breakdown. A relatively high FBP and FPP is observed. A fracture must have been created after this cycle. In the second injection cycle, oil-based mud without LCM was used to re-fracture the borehole. A much lower FBP and FPP was measured due to loss of tensile strength and removal of LCM. In the third cycle, oil-based mud with LCM was reused to strengthen the borehole. FBP and

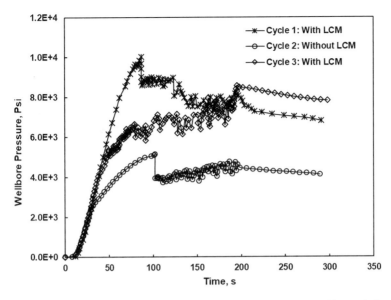

Fig. 4.19 A typical set of DEA-13 experimental data (reproduced after [6], with permission from SPE)

FPP in this cycle with LCM were about 5000 psi higher than those in the previous cycle without using LCM. As an early effort focusing on resolving lost circulation problems, the DEA-13 experimental study for the first time clearly substantiated that pressure-bearing capacity of a wellbore can be effectively enhanced by wellbore strengthening treatments.

Figure 4.20 is the result of an experimental study on remedial wellbore strengthening performed by [3]. A rock block of 6 × 6 × 6 in. with a 0.1-in.-diameter vertical borehole was used in the test. The rock sample used in this test was Grinshill sandstone. The maximum and minimum horizontal stresses applied to the block were 800 and 200 psi, respectively. Drilling mud without LCM was first injected to fracture the intact wellbore and a fracture breakdown pressure of 870 psi was achieved. Subsequently, two repeated injection cycles using the same drilling mud without LCM were conducted to test the strength of the wellbore with existing fractures. The breakdown pressure in these two cycles was about 400 psi lower than that of the intact wellbore due to the loss of tensile strength. In a final cycle, drilling mud including 30-lb/bbl graphitic LCM was used to strengthen the wellbore. Figure 4.20 shows an enhanced fracture breakdown pressure of about 1700 psi was achieved in this cycle, which is about 800 and 1200 psi higher than the breakdown pressure of the intact wellbore and fractured wellbore, respectively. This example again evidences the effectiveness of wellbore strengthening treatments.

Figure 4.21 shows a field wellbore strengthening test by [17]. The test was conducted at a depth of 3012 ft in a vertical well in the Arkoma basin, USA. A base mud free of LCM was first pumped to fracture a vertical wellbore, and an original

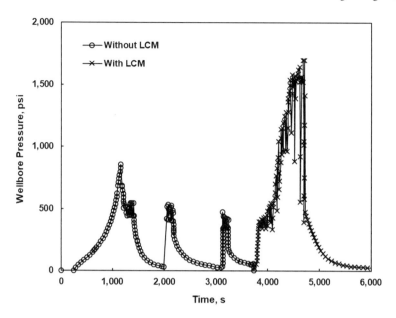

Fig. 4.20 Results of a laboratory wellbore strengthening test (reproduced after [3], with permission from SPE)

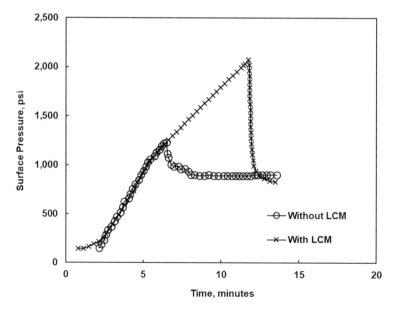

Fig. 4.21 Results of LOT field tests before and after taking remedial wellbore strengthening treatment (reproduced after [17], with permission from SPE)

breakdown pressure of 1200 psi was observed. In a following test, the base mud was replaced by a designed mud containing 80-lb/bbl LCM solids to investigate the effect of remedial wellbore strengthening. The solid curve in Fig. 4.21 shows the pressure-time curve using LCMs. A fracture (breakdown) pressure about 2050 psi was reached in this case, which is about 850 psi higher than the original state. This significant increase in fracture (breakdown) pressure clearly indicates that the fractures can be successfully bridged using remedial wellbore strengthening treatment.

References

1. Arlanoglu C, Feng Y, Podnos E, Becker E, Gray KE (2014) Finite element studies of wellbore strengthening. In: IADC/SPE drilling conference and exhibition, 4–6 March, Fort Worth, Texas, USA. https://doi.org/10.2118/168001-MS
2. Feng Y, Gray KE (2016) A parametric study for wellbore strengthening. J Nat Gas Sci Eng 30:350–363
3. Guo Q, Cook J, Way P, Ji L, Friedheim JE (2014) A comprehensive experimental study on wellbore strengthening. In: IADC/SPE drilling conference and exhibition, 4–6 March, Fort Worth, Texas, USA. https://doi.org/10.2118/167957-MS
4. Morita N, Black AD, Fuh GF (1996) Borehole breakdown pressure with drilling fluids—I. Empirical results. Int J Rock Mech Min Sci Geomech Abstr 33(1):39–51
5. Morita N, Black AD, Guh GF (1990) Theory of lost circulation pressure. In: SPE annual technical conference and exhibition, 23–26 September, New Orleans, Louisiana. https://doi.org/10.2118/20409-MS
6. Onyia EC (1994) Experimental data analysis of lost-circulation problems during drilling with oil-based mud. SPE Drilling Completion 9(1):25–31
7. Feng Y, Gray KE (2017) Review of fundamental studies on lost circulation and wellbore strengthening. J Petrol Sci Eng 152:511–522. https://doi.org/10.1016/j.petrol.2017.01.052
8. Zhong R, Miska S, Yu M (2017) Modeling of near-wellbore fracturing for wellbore strengthening. J Nat Gas Sci Eng 38:475–484
9. Feng Y, Gray KE (2016) A fracture-mechanics-based model for wellbore strengthening applications. J Nat Gas Sci Eng 29:392–400
10. Ito T, Zoback MD, Peska P (2001) Utilization of mud weights in excess of the least principal stress to stabilize wellbores: theory and practical examples. SPE Drilling Completion 16(4):221–229
11. Morita N, Fuh G-F (2012) Parametric analysis of wellbore-strengthening methods from basic rock mechanics. SPE Drilling Completion 27(2):315–327
12. van Oort E, Razavi OS (2014) Wellbore strengthening and casing smear: the common underlying mechanism. In: IADC/SPE drilling conference and exhibition, 4–6 March, Fort Worth, Texas, USA. https://doi.org/10.2118/168041-MS
13. Alberty MW, McLean MR (2004) A physical model for stress cages. In: SPE annual technical conference and exhibition, 26–29 September, Houston, Texas. https://doi.org/10.2118/90493-MS
14. Wang H, Towler BF, Soliman MY (2007) Near wellbore stress analysis and wellbore strengthening for drilling depleted formations. In: Rocky mountain oil & gas technology symposium, 16–18 April, Denver, Colorado, USA. https://doi.org/10.2118/102719-MS
15. Zoback MD (2010) Reservoir geomechanics. Cambridge University Press
16. Feng Y, Jones JF, Gray KE (2016) A review on fracture-initiation and -propagation pressures for lost circulation and wellbore strengthening. SPE Drilling Completion 31(2):134–144
17. Aston MS, Alberty MW, McLean MR, De Jong HJ, Armagost K (2004) Drilling fluids for wellbore strengthening. In: IADC/SPE drilling conference, 2–4 March, Dallas, Texas. https://doi.org/10.2118/87130-MS

18. Dupriest FE (2005) Fracture closure stress (FCS) and lost returns practices. In: SPE/IADC drilling conference, 23–25 February, Amsterdam, Netherlands. https://doi.org/10.2118/92192-MS
19. van Oort E, Friedheim JE, Pierce T, Lee J (2011) Avoiding Losses in depleted and weak zones by constantly strengthening wellbores. SPE Drilling Completion 26(4):519–530
20. Growcock FB, Kaageson-Loe N, Friedheim J, Sanders MW, Bruton J (2009) Wellbore stability, stabilization and strengthening. In: Offshore mediterranean conference and exhibition, 25–27 March, Ravenna, Italy
21. Aston MS, Alberty MW, Duncum SD, Bruton JR, Friedheim JE, Sanders MW (2007) A new treatment for wellbore strengthening in shale. In: SPE annual technical conference and exhibition, 11–14 November, Anaheim, California, USA. https://doi.org/10.2118/110713-MS
22. Dupriest FE, Smith MV, Zeilinger SC, Shoykhet N (2008) Method to eliminate lost returns and build integrity continuously with high-filtration-rate fluid. in: IADC/SPE drilling conference, 4–6 March, Orlando, Florida, USA. https://doi.org/10.2118/112656-MS
23. Feng Y, Arlanoglu C, Podnos E, Becker E, Gray KE (2015) Finite-Element studies of hoop-stress enhancement for wellbore strengthening. SPE Drilling Completion 30(1):38–51
24. Wang H, Soliman MY, Towler BF (2009) Investigation of factors for strengthening a wellbore by propping fractures. SPE Drilling Completion 24(3):441–451
25. Wang H, Towler BF, Soliman MY (2007) Fractured wellbore stress analysis: sealing cracks to strengthen a wellbore. in: SPE/IADC drilling conference, 20–22 February, Amsterdam, The Netherlands. https://doi.org/10.2118/104947-MS

Chapter 5
Lost Circulation Materials

Abstract Numerous LCMs have been proposed for curing lost circulation. However, due to the complexity of formation properties and fluid loss mechanisms, there is not a universal LCM suitable to all types of losses. The selection of LCMs is highly dependent on the formation types, fluid loss mechanisms (e.g. loss into pores, natural fractures, induced fractures, and vugs), and the loss severity. This chapter gives a brief overview of the LCMs that can be used to mitigate fluid losses in high-permeability formations, low-permeability shales, and vugular/fractured carbonates.

5.1 Introduction

Lost circulation materials (LCMs) are the materials used to bridge and seal the flow channels (pores, fractures, vugs) through which the drilling fluids lost into the formation. Based on their appearances, LCMs are usually divided into four categories [1, 2]:

- Granular: synthetic graphite, calcium carbonate, nut shells, gilsonite, asphalt, etc.,
- Fibrous: cellulose fibers, nylon, mineral fibers, cedar bark, shredded paper, etc.,
- Flaky: mica, cellophane, etc.,
- Blends of two or more of these materials.

Granular LCMs are used to form a bridge within the pores or fractures to prevent fluid loss. The have relatively high stiffness and are available in a wide range of particle size distribution (PSD). The design of PSD of granular materials is critical for wellbore strengthening. Generally, large particles are used to bridge the fractures, while sufficient small particles should also be included to fill the spaces between the large ones to form a low-permeability seal. So LCMs with a broad PSD usually perform better than those with a narrow PSD [3].

Fibers are flexible and slender materials. Due to their very low stiffness, using fibers alone in wellbore strengthening treatments may not be able to create an effective bridge. Therefore, fibers are usually used together with granular LCMs. A major function of fibrous LCMs is to create a mat-like bridge which serves as a filtration medium for granular materials to deposit and to form a tight seal, eventually [4].

© The Author(s) 2018
Y. Feng and K. E. Gray, *Lost Circulation and Wellbore Strengthening*,
SpringerBriefs in Petroleum Geoscience & Engineering,
https://doi.org/10.1007/978-3-319-89435-5_5

Using fibers in combination with granular materials can also improve the mechanical properties and seepage characteristic of the bridge.

Flaky materials are thin and flat in shape. Flaky LCMs can establish seal over many permeable formations to stop mud loss [5]. However, due to their relatively low strength, flaky LCMs are more likely to be destroyed during placement. An experimental study shows that a flaky LCM is not effective in sealing lost circulation fractures when used alone [6].

A blend of different types of LCMs can yield better performance by leveraging the property and advantage of each material. A commonly used prescription for curing lost circulation in permeable and fractured zones is a blend of sized calcium carbonate and resilient synthetic graphite, sometimes supplemented with fibers [7].

According to the settability of the materials, LCMs can be classified into:

- Settable materials: cement, resin, cross-linked material, etc.,
- Non-settable materials,
- Blends of settable and non-settable materials.

Settable materials solidify chemically in the pores, fractures, or vugs to block the fluid loss channels. Non-settable LCMs consolidate and pack in the openings to prevent fluid loss, which is usually dependent on fluid leak-off.

Based on the requirement of fluid loss for forming the seal within the fracture, LCM treatments can be classified into [7]:

- Low-fluid-loss treatment,
- High-fluid-loss treatment.

Low-fluid-loss treatments are effective when the fracture can be sealed rapidly by using materials such as cement, cross-linked polymers, and particulates that pack quickly. High-fluid-loss treatments form seals within the fracture as the materials deposit in the fracture by a process of de-fluidization. Fluid leak-off is required for this process, and thus this approach is effective only in high-permeability formations or fractured formations exhibiting high fluid loss [7].

No universal LCM can treat all lost circulation events due to the complexity of the problem. The selection of LCMs are usually based on the mechanisms and severity of fluid losses. The fluid loss mechanisms mainly include loss into:

- pores of high-permeability rocks,
- natural fractures,
- drilling-induced fractures,
- vugs/caverns.

The severity of fluid loss is generally classified into four grades based on the loss rate [8]:

- seepage loss: 1–10 bbl/h,
- partial loss: 10–100 bbl/h,
- severe loss: >100 bbl/h,
- total loss: no returns.

Both the mechanisms and severity of fluid loss are determined by the properties of the formation. In this chapter, a brief overview of LCMs for different types of formations is presented.

5.2 Lost Circulation Materials for High-Permeability Formations

High-permeability formations where lost circulation events are usually encountered include unconsolidated sands, pea gravels, and depleted reservoirs [9]. Large pore spaces and drilling-induced fractures (in depleted zones) are typical channels for fluid loss in these formations. The loss severity can range from seepage loss to total loss dependent on the size of the openings [8].

It is relatively easier to form a bridge or seal inside a fracture in permeable formations thanks to the high fluid leak-off rate [10–13]. Therefore, non-settable LCMs should be first considered in high-permeability formations unless very large loss occurs. The non-settable materials can be granular, fibrous, and flaky materials. A most commonly used LCM to create a seal is calcium carbonate. Experimental results have shown that blending calcium carbonate with synthetic graphite can significantly improve the sealing performance because of the unique resilient property of the graphite [14, 15]. A 3/2 graphite to calcium carbonate ratio shows much less fluid loss compare with using just one type of the materials [16].

High-fluid-loss treatment can be used in high-permeability formations because the fluid leak-off can ensure the creation of a tight seal in the fracture [17, 18]. It is important to minimize the fracture surface contamination in order to facilitate leak-off and material deposition in the fracture. Therefore, LCM pills are suggested for high-fluid-loss treatments to minimize fracture contamination, rather than using LCMs in the whole drilling mud. It may be difficult to seal fractures with aperture larger than 2 mm using high-fluid-loss treatment because the large flow rate in the fracture may impede the form of the seal [7].

Properly sized LCMs should be used to plug fractures in high-permeability formations based on the size of the fractures. The fracture size (aperture) can be estimated from rock mechanical modeling [19], such as the models presented in Chap. 3. Large enough LCM particles are required to form a bridge in the fracture and, on the other hand, there should be plenty of fine particles sealing the spaces between the large particles. The ideal packing theory can be used to optimize the PSD of LCMs, which suggests that the D90 value of the PSD should be equal to the size of the fracture opening [20, 21]. Another criterion suggests that D50 of the PSD should match the median opening of the fracture [3, 8]. A multimodal PSD criterion is also used for LCM selection and shows better results than a normal PSD distribution [22].

Fibers and cement have been recommended for curing severe losses in high-permeability formations [3, 9, 23]. Cross-linked materials are also used for lost circulation control in depleted sands and unconsolidated formations [24, 25].

However, many high-permeability formations are pay zones. Reservoir-friendly LCMs are needed in such zones to avoid formation damage. An advantage of the most commonly used calcium carbonate is its acid-soluble feature, so it is reservoir-friendly. Traditional fibers and cement are usually not reservoir-friendly because they are not removable by acid or other treatments. To this end, acid-soluble fibers and self-degradable fibers have been developed for curing lost circulation in pay zones [26–28]. Acid-soluble cements are also proposed and applied to pay zones [29, 30]. Such cement can form seals with comparable strength to traditional cement, but the seals can be removed later with acid to minimize formation damage. Some acid-soluble crosslinked gels are also reported for severe lost circulation in depleted reservoirs [31].

5.3 Lost Circulation Materials for Low-Permeability Shales

In low-permeability shales, fluid losses are mainly caused by natural fractures or drilling-induced fractures. The loss can vary from partial loss caused by small natural or induced fractures to total loss caused by large fractures.

As described in Sect. 1.3.2 "Physical Models of Wellbore Strengthening", all the current wellbore strengthening theories are based on the presence of sufficient fluid leak-off from the fracture to form an effective seal. A critical issue for wellbore strengthening in low-permeability shales is the limited leak-off of the trapped fluid within the fracture. LCMs may not deposit and de-fluidize in the fracture. As a result, the wellbore pressure can readily transmit to the fracture tip and drive fracture propagation [12, 13, 32]. During temporarily drop in wellbore pressure associated with normal drilling stops, the loose LCM deposition in the fracture will be carried back into the wellbore with fluid back flow, destroying the seal of the fracture. Contrarily, in a permeable formation, rapid leak-off of the trapped fluid in the fracture permits the development of a much stronger seal. In addition, the bleed-off of the trapped pressure will prevent fluid back flow during regular drilling stops [11, 33]. Therefore, the key to cure losses in shales is to develop a technology to create a tight seal in the fracture that does not depend on fluid leak-off [11].

An effective approach is using settable materials. A settable treatment pill has been proven effective for strengthening shales [11]. The pill is formulated with the ingredients of base oil, gelling agent, cross-linked polymers, hardening agent, viscosifier, and bridging particles. The pill sets with time and the setting time can be controlled by modifying the concentrations of the additives. A field trial conducted in the US Arkoma basin shows that this settable pill raised the fracture reopening pressure and fracture breakdown pressure of a formation at 4020 ft by 550 and 150 psi, respectively.

Lost circulation in sub-salt shales has always been a challenging drilling problem [34–36]. The sub-salt thief zone is usually highly fractured, with complex fracture networks. Sized solids, gunk, and cement have all been applied to solve losses in sub-salt zones. A pill blend of cross-linking polymers and fibers was successfully used for controlling lost circulation in Gulf of Mexico sub-salt thief zones [37, 38]. The pill sets with time and produces a rubbery, spongy, and ductile seal in the fracture.

The design of LCM PSD for sealing fractures in shales shares the same criteria for sealing fractures in permeable formations. The particles should be able to enter the fracture and deposit close to the wellbore. Sealing the fracture near the wellbore can maximize the wellbore strengthening effect and minimize the loss volume. This requires that the particle size should not be very small, otherwise the particles will move deep into the fracture. On the other hand, the particles should not be too large, otherwise they will not enter the fracture and will deposit on the wellbore wall which can be destroyed easily by drilling operations.

In recent years, nanoparticles technology has gained an increasing interest in overcoming lost circulation and wellbore instability problems in shales [39–43]. Nanoparticles are the particles with a size in the range of 1–100 nm. They are much smaller than the fine particles used in traditional muds (about a magnitude smaller than the size of bentonite). Nanoparticles can enter the small openings of shales and form an effective seal to the micro cracks in shales. Plugging of shale pores by nanoparticles provides a powerful new solution for lost circulation and wellbore instability. Another important merit of nanoparticles is their ability to modify the properties of LCM slurries/pills. A cross-linked nanocomposite gel pill has been proposed for combating severe fluid losses [24]. The pill consists of four major components: an aqueous viscous polymer base, a cross-linking agent, swelling cross-linked grains, and colloidal particles. This pill is a settable system which gels to seal off loss zones.

5.4   Lost Circulation Materials for Carbonate Formations

Lost circulation in carbonate formations are usually caused by the presence of vugs and fractures in the formation. Losses in vulgar and fractured carbonate formations can be severe to total losses, which are among the most challenging lost circulation problems. Curing such losses can be difficult with particulate LCM treatments.

Cement is a commonly used and effective material to combat losses in carbonate formations because cement can easily enter the vugs or fractures and obstruct them quickly. However, cement can also plug formation pores and make the subsequent production impossible if the carbonate formation is a pay zone. For solving this problem, reservoir-friendly, acid-soluble cements have been developed to cure fluid loss in carbonate pay zones [9, 29, 30]. Customized thixotropic and ultra-thixotropic cement slurries containing falkes, mica, calcium carbonate (for improving mechanical properties of the seal), and unique spacer and surfactant were also applied to control losses [23].

Crosslinking gels can also be effective for sealing fractures and vugs in carbonate formations. Crosslinking polymers produce a gel structure in the openings by the formation of crosslinked bonds between the polymer chains [24]. Acid-soluble crosslinking gels were also developed for using in pay zones [31].

LCM pills consisting of particle materials and fibers were also used for fixing lost circulation in vugular and fractured formations [44]. Adding fibers provides a bridge structure as a support for the particulate materials and promotes an increase in the mechanical strength of the seal.

Last but not least, no matter the lost circulation scenario, it is always a wise choice to avoid lost circulation by carefully design of the mud density and rheology, before considering using LCMs. If this fails to avoid losses, non-settable LCMs, such as calcium carbonate and fibers, should be considered first. If such materials still do not solve the problem, settable LCMs, such as cement, cross-linking polymers, and gels, can be used.

References

1. Canson BE (1985) Lost circulation treatments for naturally fractured, vugular, or cavernous formations. In: SPE/IADC drilling conference, 5–8 March, New Orleans, Louisiana. https://doi.org/10.2118/13440-MS
2. White RJ (1956) Lost-circulation materials and their evaluation. In: Drilling and production practice. Am Pet Inst, New York
3. Lavrov A (2016) Lost circulation: mechanisms and solutions, 1st edn. Gulf Professional Publishing, Cambridge
4. Alsaba M, Nygaard R, Hareland G, Contreras O (2014) Review of lost circulation materials and treatments with an updated classification. In: AADE national technical conference and exhibition, Houston, pp 15–16
5. Nayberg TM (1987) Laboratory study of lost circulation materials for use in both oil-based and water-based drilling muds. SPE Drilling Eng 2(03):229–236. https://doi.org/10.2118/14723-PA
6. Whitfill DL, Hemphill T (2003) All lost-circulation materials and systems are not created equal. In SPE annual technical conference and exhibition, 5–8 October, Denver, Colorado. https://doi.org/10.2118/84319-MS
7. Growcock FB, Kaageson-Loe N, Friedheim J, Sanders MW, Bruton J (2009) Wellbore stability, stabilization and strengthening. In: Offshore mediterranean conference and exhibition, 25–27 March, Ravenna, Italy
8. Nelson EB (1990) Well cementing, vol 28. Newnes
9. Luzardo J et al (2015) Alternative lost circulation material for depleted reservoirs. In: OTC Brasil, 27–29 October, Rio de Janeiro, Brazil. https://doi.org/10.4043/26188-MS
10. Alberty MW, McLean MR (2004) A physical model for stress cages. In: SPE annual technical conference and exhibition, 26–29 September, Houston, Texas. https://doi.org/10.2118/90493-MS
11. Aston MS, Alberty MW, Duncum SD, Bruton JR, Friedheim JE, Sanders MW (2007) A new treatment for wellbore strengthening in shale. In: SPE annual technical conference and exhibition, 11–14 November, Anaheim, California, USA. https://doi.org/10.2118/110713-MS
12. Feng Y, Jones JF, Gray KE (2016) A review on fracture-initiation and -propagation pressures for lost circulation and wellbore strengthening. SPE Drilling Completion 31(02):134–144
13. Feng Y, Gray KE (2017) Review of fundamental studies on lost circulation and wellbore strengthening. J Petrol Sci Eng 152:511–522
14. Savari S, Whitfill DL, Kumar A (2012) Resilient lost circulation material (LCM): a significant factor in effective wellbore strengthening. In SPE deepwater drilling and completions conference, 20–21 June, Galveston, Texas, USA. https://doi.org/10.2118/153154-MS

15. Whitfill D (2008) Lost circulation material selection, particle size distribution and fracture modeling with fracture simulation software. In: IADC/SPE Asia Pacific drilling technology conference and exhibition, 25–27 August, Jakarta, Indonesia. https://doi.org/10.2118/115039-MS

16. Fekete PO, Dosunmu A, Kuerunwa A, Ekeinde EB, Chimaroke A, Baridor OS (2013) Wellbore stability management in depleted and low pressure reservoirs. In: SPE Nigeria annual international conference and exhibition, 5–7 August, Lagos, Nigeria. https://doi.org/10.2118/167543-MS

17. Sanders MW, Scorsone JT, Friedheim JE (2010) High-fluid-loss, high-strength lost circulation treatments. In: SPE deepwater drilling and completions conference, 5–6 October, Galveston, Texas, USA. https://doi.org/10.2118/135472-MS

18. Wang H, Sweatman RE, Engelman RE, Deeg WF, Whitfill DL (2005) The key to successfully applying today's lost circulation solutions. In: SPE annual technical conference and exhibition, 9–12 October, Dallas, Texas. https://doi.org/10.2118/95895-MS

19. Feng Y, Gray KE (2018) Modeling Lost circulation through drilling-induced fractures. SPE J 23(01):205–223

20. Dick MA, Heinz TJ, Svoboda CF, Aston M (2000) Optimizing the selection of bridging particles for reservoir drilling fluids. In: SPE international symposium on formation damage control, 23–24 February, Lafayette, Louisiana. https://doi.org/10.2118/58793-MS

21. Kageson-Loe NM et al (2009) Particulate-based loss-prevention material–the secrets of fracture sealing revealed! SPE Drilling Completion 24(04):581–589

22. Savari S, Whitfill DL (2015) Managing losses in naturally fractured formations: sometimes nano is too small. In: SPE/IADC drilling conference and exhibition, 17–19 March, London, England, UK. https://doi.org/10.2118/173062-MS

23. Fidan E, Babadagli T, Kuru E (2004) Use of cement as lost-circulation material: best practices. In: Canadian international petroleum conference, 8–10 June, Calgary, Alberta. https://doi.org/10.2118/2004-090

24. Lecolier E, Herzhaft B, Rousseau L, Neau L, Quillien B, Kieffer J (2005) Development of a nanocomposite gel for lost circulation treatment. In: SPE European formation damage conference, 25–27 May, Sheveningen, The Netherlands. https://doi.org/10.2118/94686-MS

25. Wang H et al (2008) Best practice in understanding and managing lost circulation challenges. SPE Drilling Completion 23(02):168–175

26. Droger N et al (2014) Degradable fiber Pill for lost circulation in fractured reservoir sections. In: IADC/SPE drilling conference and exhibition, 4–6 March, Fort Worth, Texas, USA. https://doi.org/10.2118/168024-MS

27. Ghassemzadeh J (2013) Lost circulation material for oilfield use. U.S. Patent No. 8,404,622

28. Halliday WS, Clapper DK, Jarrett M, Carr M (2004) Acid soluble, high fluid loss pill for lost circulation. U.S. Patent No. 6,790,812

29. Al-yami AS, Al-Ateeq A, Wagle VB, Alabdullatif ZA, Qahtani M (2014) New developed acid soluble cement and sodium silicate gel to cure lost circulation zones. In: Abu Dhabi international petroleum exhibition and conference, 10–13 November, Abu Dhabi, UAE. https://doi.org/10.2118/172020-MS

30. Vinson EF, Totten PL, Middaugh RL (1992) Acid removable cement system helps lost circulation in productive zones. In: SPE/IADC drilling conference, 18–21 February, New Orleans, Louisiana. https://doi.org/10.2118/23929-MS

31. Suyan KM, Sharma V, Jain VK (2009) An innovative material for severe lost circulation control in depleted formations. In: Middle East drilling technology conference & exhibition, 26–28 October, Manama, Bahrain. https://doi.org/10.2118/125693-MS

32. Feng Y, Jones JF, Gray KE (2015) Pump-in and flow-back test for determination of fracture parameters and in-situ stresses. In: AADE 2015 national technical conference and exhibition, 8–9 April, San Antonio, Texas, USA

33. Dupriest FE (2005) Fracture closure stress (FCS) and lost returns practices. In: SPE/IADC drilling conference, 23–25 February, Amsterdam, Netherlands. https://doi.org/10.2118/92192-MS

34. Fredrich JT, Engler BP, Smith JA, Onyia EC, Tolman D (2007) Pre-drill estimation of sub-salt fracture gradient: analysis of the Spa prospect to validate non-linear finite element stress analyses. In: SPE/IADC drilling conference, 20–22 February, Amsterdam, The Netherlands. https://doi.org/10.2118/105763-MS

35. Power D, Ivan CD, Brooks SW (2003) The top 10 lost circulation concerns in deepwater drilling. In: SPE Latin American and Caribbean petroleum engineering conference, 27–30 April, Port-of-Spain, Trinidad and Tobago. https://doi.org/10.2118/81133-MS

36. Willson SM, Fredrich JT (2005) Geomechanics considerations for through-and near-salt well design. In: SPE annual technical conference and exhibition, 9–12 October, Dallas, Texas. https://doi.org/10.2118/95621-MS

37. Caughron DE et al (2002) Unique crosslinking pill in tandem with fracture prediction model cures circulation losses in deepwater Gulf of Mexico. In: IADC/SPE drilling conference, 26–28 February, Dallas, Texas. https://doi.org/10.2118/74518-MS

38. Ferras M, Galal M, Power D (2002) Lost circulation solutions for severe sub-salt thief zones. In: AADE 2002 Technology Conference Drilling & Completion Fluids and Waste Management, 2–3 April, Houston, Texas, USA

39. Contreras O, Hareland G, Husein M, Nygaard R, Al-saba MT (2014) Experimental investigation on wellbore strengthening in shales by means of nanoparticle-based drilling fluids. In: SPE annual technical conference and exhibition, 27–29 October, Amsterdam, The Netherlands. https://doi.org/10.2118/170589-MS

40. Friedheim JE, Young S, De Stefano G, Lee J, Guo Q (2012) Nanotechnology for oilfield applications-hype or reality? In: SPE international oilfield nanotechnology conference and exhibition, 12–14 June, Noordwijk, The Netherlands. https://doi.org/10.2118/157032-MS

41. Hoelscher KP, De Stefano G, Riley M, Young S (2012) Application of nanotechnology in drilling fluids. In: SPE international oilfield nanotechnology conference and exhibition, 12–14 June, Noordwijk, The Netherlands. https://doi.org/10.2118/157031-MS

42. Li G, Zhang J, Hou Y (2012) Nanotechnology to improve sealing ability of drilling fluids for shale with micro-cracks during drilling. In: SPE international oilfield nanotechnology conference and exhibition, 12–14 June, Noordwijk, The Netherlands. https://doi.org/10.2118/156997-MS

43. Nwaoji CO (2012) Wellbore strengthening- nano-particle drilling fluid experimental design using hydraulic fracture apparatus. Thesis, University of Calgary

44. Kumar A, Savari S, Whitfill D, Jamison D (2011) Application of fiber laden pill for controlling lost circulation in natural fractures. In: AADE 2011 national technical conference and exhibition, 12–14 April, Houston, Texas, USA

Chapter 6
Recommendations for Future Work

Abstract Although there have been a lot of studies dedicated to lost circulation control and great successes have been achieved, the problem continues, with the increasing drilling operations in more complicated environments. Knowledge gaps and unsolved problems still exist and the need for further studies remains. A few recommendations for future work are presented in this chapter.

6.1 Recommendations for Future Work

This short monograph is a brief study on lost circulation and wellbore strengthening. Several analytical and numerical models were developed to model dynamic fluid loss while drilling, preventive wellbore strengthening based on bridging/plugging lost circulation fractures. Unsolved problems still exist as more complicated drilling conditions are encountered. This last chapter suggests some recommendations for future research related to lost circulation and wellbore strengthening:

- It is highly recommended to consider the effect of pre-existing fractures on the wellbore wall in modeling lost circulation. Lost circulation fractures may not propagate perfectly along the direction of maximum horizontal stress under this condition.
- Thermal effects should be considered in modeling studies of lost circulation and wellbore strengthening, especially for HPHT wells and geothermal wells.
- Studies can be extended to lost circulation based on forming a layer of tight mudcake on the wellbore wall. Dynamic mud filtration should be investigated to understand time-dependent developments of both external and internal mudcake.
- Advanced numerical models with capabilities of simulating transportation and deposition of LCMs in the lost circulation fractures would be very useful for modeling dynamic fracture bridging/plugging process in wellbore strengthening.
- For better application of wellbore strengthening techniques, it is important to accurately and quickly estimate/measure the geometry of lost circulation fractures during drilling for selecting/adjusting the size distribution of LCMs in real time.

© The Author(s) 2018   81
Y. Feng and K. E. Gray, *Lost Circulation and Wellbore Strengthening*,
SpringerBriefs in Petroleum Geoscience & Engineering,
https://doi.org/10.1007/978-3-319-89435-5_6

Improved or new logging while drilling techniques are needed for acquiring better knowledge of drilling-induced or pre-exiting natural fractures on the wellbore wall.

- Current wellbore strengthening studies mainly focus on addressing fluid loss through hydraulically induced fractures in sand/shale formations. Severe losses are also commonly encountered in carbonate formations with vugs, cavities, and large fractures. More efforts are needed to address lost circulation in carbonates.
- Lost circulation events in anisotropic/heterogeneous formations with complex lithology, stress, and pressure profiles have posed great challenges to drilling and wellbore strengthening. Efforts on addressing problems in such formations are needed.
- More advanced environment-friendly LCMs and reservoir-friendly LCMs need development.